Decarbonizing Rural Buildings and Rural Energy System

Building Energy Research Center Tsinghua University

Decarbonizing Rural Buildings and Rural Energy System

China Building Energy and Emission Yearbook 2024

 Springer

Building Energy Research Center Tsinghua University
Beijing, China

ISBN 978-981-97-9119-4   ISBN 978-981-97-9120-0 (eBook)
https://doi.org/10.1007/978-981-97-9120-0

This work was supported by Tsinghua University.

© Building Energy Research Center, Tsinghua University 2025. This book is an open access publication.

Open Access This book is licensed under the terms of the Creative Commons Attribution-NonCommercial-NoDerivatives 4.0 International License (http://creativecommons.org/licenses/by-nc-nd/4.0/), which permits any noncommercial use, sharing, distribution and reproduction in any medium or format, as long as you give appropriate credit to the original author(s) and the source, provide a link to the Creative Commons license and indicate if you modified the licensed material. You do not have permission under this license to share adapted material derived from this book or parts of it.

The images or other third party material in this book are included in the book's Creative Commons license, unless indicated otherwise in a credit line to the material. If material is not included in the book's Creative Commons license and your intended use is not permitted by statutory regulation or exceeds the permitted use, you will need to obtain permission directly from the copyright holder.

This work is subject to copyright. All commercial rights are reserved by the author(s), whether the whole or part of the material is concerned, specifically the rights of reprinting, reuse of illustrations, recitation, broadcasting, reproduction on microfilms or in any other physical way, and transmission or information storage and retrieval, electronic adaptation, computer software, or by similar or dissimilar methodology now known or hereafter developed. Regarding these commercial rights a non-exclusive license has been granted to the publisher.

The use of general descriptive names, registered names, trademarks, service marks, etc. in this publication does not imply, even in the absence of a specific statement, that such names are exempt from the relevant protective laws and regulations and therefore free for general use.

The publisher, the authors and the editors are safe to assume that the advice and information in this book are believed to be true and accurate at the date of publication. Neither the publisher nor the authors or the editors give a warranty, expressed or implied, with respect to the material contained herein or for any errors or omissions that may have been made. The publisher remains neutral with regard to jurisdictional claims in published maps and institutional affiliations.

This Springer imprint is published by the registered company Springer Nature Singapore Pte Ltd.
The registered company address is: 152 Beach Road, #21-01/04 Gateway East, Singapore 189721, Singapore

If disposing of this product, please recycle the paper.

Preface

Since the introduction of the carbon peaking and carbon neutrality strategy by the Chinese government in 2020, the main task for achieving carbon neutrality has been to transform the current carbon-based energy structure into a zero-carbon energy structure. Rural areas are pivotal in this transformation, given their ample space and renewable resources, which can support the exploration of wind, solar, hydro, and biomass energy. Therefore, this report comprehensively discusses the issues related to the new rural energy system, exploring how to meet the current needs of rural revitalization and energy demand, comprehensively achieve future carbon neutrality goals, and gradually transition from the existing system to a new one while ensuring social development, resident livelihoods, and economic growth.

After more than four years of continuous research led by Prof. Xudong Yang of the Building Energy Research Center of Tsinghua University, solid ideas have been formulated regarding the structure, characteristics, and operation modes of the rural energy system under the carbon neutrality vision in China. Additionally, initial clarification has been made on how to ensure the current energy demands for rural area, as well as the growing needs, and how to transform the existing energy supply and provide system step by step, eventually transitioning to a new energy system that meets the requirements of carbon neutrality. This report provides a comprehensive summary of the research results mentioned above, describing the current status of energy use situation and energy saving paths of rural buildings, and pointing out the key solution to China's energy revolution. It also introduces the challenges faced by the rural energy system in achieving carbon neutrality and elucidates the low-carbon energy structure primarily based on photovoltaic and biomass fuels which are abundant in rural area. Consequently, under the carbon neutrality vision, the rural areas are expected to undertake half of the photovoltaic power generation tasks, become the main source of zero-carbon fuels, and undertake 25% of the power storage and regulation tasks, thus shift the current coal-fired power plants to a large proportion of wind and photovoltaic power. A new zero-carbon power system could be build based on renewable energy, by shifting from a centralized approach to a combination of centralized and distributed energy system. It can also serve as an experimental and demonstration site for new power system policy mechanisms, gradually expanding to a wider area.

Regarding zero-carbon fuels, the report discusses fuels processed from biomass resources and synthetic fuels like ammonia produced from hydrogen electrolysis using green electricity. It highlights the challenges in commercializing biomass energy due to technological, economic, and policy hurdles, advocating for effective policy mechanisms to promote their development.

Overall, the new energy system is anticipated to drive the transformation of the rural economy, turning rural areas from energy consumers into energy providers. This shift aims to reduce environmental pollution while boosting economic growth.

Yi Jiang, Ph.D.
Academician of CAE
Professor and Director, Tsinghua
Building Energy Research Center
Beijing, China

Acknowledgements

This publication was prepared by the Building Energy Research Center (BERC) of Tsinghua University.

The lead authors were Prof. Yi Jiang, Prof. Xudong Yang, Dr. Ming Shan, Dr. Shan Hu.

Other authors also contributed important sections for this report, and they were as follows:

Dr. Yang Zhang (Chaps. 1, 2)
Ziyi Yang (Chaps. 1, 2)
Zhenghua Wang (Chaps. 1, 2)
Dr. Xingli Ding (Sects. 3.4.1, 3.4.3)
Xin He (Sect. 3.4.2)
Dr. Yazhou Nie (Chap. 4)
Dr. Tao Sun (Chap. 5)
Dr. Yuan Zhi (Sects. 6.1, 6.2.2, 6.3, 7.2, 7.3)
Ding Gao (Sects. 6.2.1, 6.2.3, 6.2.4, 6.3)
Xing Rong (Sect. 7.1)
Zhigang Zhao (Sects. 7.4, 7.5)
Heping Liu (Sects. 7.4, 7.5)
Jinrong Yuan (Sects. 7.4, 7.5)
Dr. Xiaomeng Chen (Sect. 7.6)
Prof. Yuping Dong (Sect. 7.7)
Prof. Xiujin Li (Sect. 7.8)

The report was final compiled by Xingli Ding.

The research of this report was supported by Chinese Academy of Engineering Strategic Research and Consulting Program "Research on Low Carbon Development Strategy of Urban and Rural Energy Supply System" (Grant No. 2023-XBZD-07), and Natural Science Foundation of China (Grant No. 72261147760).

Executive Summary

Building Construction and Building Embodied Energy Use and Emissions in China Have Been Decreasing Since 2016

In 2022, China reached a population of 1412 million, with the urbanization rate rising from 37.7% in 2001 to 65.2% in 2022. Rapid urbanization drives continuous development of the building construction. In 2022, the total building stock in China reached 69.6 billion m^2, including 31.8 billion m^2 of urban residential buildings, 22.4 billion m^2 of rural residential buildings, 15.4 billion m^2 of public and commercial (P&C) buildings. The floor area for northern urban heating is 16.7 billion m^2.

The embodied energy use and embodied CO_2 emissions of China's civil buildings have been decreasing since 2016. In 2022, the embodied energy use of civil buildings in China is 0.5 Gtce in 2022, and embodied carbon emissions of civil building construction is 1.5 GtCO_2. Adequate planning and speed of construction of the entire building stock, a shift from new construction to maintenance and renovation of existing buildings, low-carbon building materials and construction methods are important issues for the low-carbon transition of China's building sector.

Energy Use and Carbon Emission Intensity of Building Operation in China Continues to Grow Steadily, But Remains Lower Compared to Developed Countries

In 2022, the total commercial energy consumption of building operation in China was 1.12 Gtce, and the total carbon emissions during building operation were 2.2 GtCO_2. The electricity consumption of building operation exceeded 2.3 PWh in 2022, leading to a corresponding increase in the indirect carbon emission from electricity use to 1.26 GtCO_2. On the other hand, the indirect carbon emission from heating during building operation was 440 million tons of CO_2, which showed a gradual but steady growth trend.

China's building operation energy consumption per capita and per square meter remains significantly lower than US, Canada, Europe, Japan, and South Korea. Specifically, China's equivalent electricity consumption per capita of building operation was approximately one-fifth of that in US and Canada, and half of that in Japan and South Korea. Similarly, the equivalent electricity consumption per square meter of building operation in China was one-third of that in Canada and half of that in US, Europe, Japan, and South Korea. As a result, China's building carbon emissions per capita and per square meter are lower than those in most developed countries.

Rural Areas in China Have a Significant Demand for Clean Heating Transformation, Making Them a Focal Point for Low-Carbon Energy Systems

Currently, the total energy consumption of rural building is 253 million tce. The consumption of commercial coal, LPG, electricity, and natural gas is 95 million tons (67 million tce), 8.07 million tons (14 million tce), 348.1 billion kWh (105 million tce), and 12.4 billion m^3 (15 million tce) respectively, accounting for 79.4% of total. The non-commercial energy, i.e., biomass (including firewood and straw), was 94 million tons (52 million tce), accounting for 20.6% of total.

The effect of Clean heating: The consumption of bulk coal in rural areas has decreased from 197 million tons in 2014 to 95 million tons in 2022, down by about 51.8%. The Beijing-Tianjin-Hebei region and its surrounding areas account for 49% and 82% of China's $PM_{2.5}$ and CO_2 emission reductions caused by domestic energy consumption in rural areas, respectively.

The decrease of non-commercial energy: The biomass energy based on straw and firewood decreased by about 87 million tons in 2022, the proportion of which fell from 31.2% to 20.6%.

Energy saving path in rural areas: Building form, occupant type, and income level, as well as the actual building space usage pattern, thermal comfort characteristics, and heating demand of RRB should be considered when realizing energy saving. The heating system and equipment of "partial-time, partial-space", featuring room-specific installation and flexible regulation, can better match the actual demand characteristics of RRB in northern China. Targeted renovation scheme of building envelop in RRB can achieve a reasonable energy-saving effect, and even optimize the initial investment in renovation.

Energy Revolution Initialized in Rural Areas

The primary objective of China's energy revolution is to establish a zero-carbon power system and a zero-carbon fuel production, supply, and marketing infrastructure. Given the abundant spatial and biomass resources in rural areas, they will transition from being energy consumers to becoming energy suppliers in the future zero-carbon energy framework. The rural areas will lead in achieving carbon neutrality and play a pivotal role in shaping the zero-carbon energy system, thus making significant contributions. Consequently, the development of a rural new energy system becomes an integral component of China's overall energy transitions.

The rural energy system is mainly designed as a new PEDF distributed system underlain by a rooftop PV system, and is also meant to increase the proportion of commercialized biomass energy products. As estimated, the current rooftop space in rural areas is enough to install about 2 billion kW PV capacity and generate 2.9 PWh every year. Therefore, fully exploiting the rural rooftop space resource will provide the energy required for living, production, and transportation, and the electricity generated will supersede coal, oil, gas, and biomass, thus realizing a complete zero-carbon economy in rural areas. The rural areas will be the most important base of zero-carbon fuels, and the biomass energy from them will account for over 70% of the future energy demand. Under ideal conditions, the rural areas will output 2.3 PWh of zero-carbon electricity and 740 million tce of zero-carbon fuel in 2060.

Potential and Characteristics of Rural Distributed Zero-Carbon Energy in China

The rural areas in China possess abundant distributed zero-carbon energy in the form of rooftop PV, biomass, and rural hydropower, but such resources have not been adequately exploited. Under the carbon neutrality goal, the rural areas should make greater efforts to develop the aforesaid resources, so as to change the rural areas from consumers of fossil energy to producers of zero-carbon energy, to contribute to the carbon neutrality endeavor of China, improve the living quality and economic income of the rural dwellers and to improve the environmental quality.

The potential total installed capacity of rural rooftop PV systems: By using satellite images, the corrected U-net neural network model has been created and applied. With consideration of PV panel installation methods, the potential total installed capacity of rural rooftop PV systems in China is calculated as 2 billion kW and the annual power generation potential is 2.9 PWh.

The biomass resources: The annual total quantity of rural biomass energy in China is about equivalent to 736 million tce. The crop straws account for about 50.6%.

The rural hydropower: The current installed capacity of the rural hydropower projects in China is about 81,338,000 kW and the annual energy production is 242.37

billion kWh. In terms of resource potential, the potential installed capacity of the rural hydropower projects in China is about 128 million kW and the annual energy production potential is about 535 billion kWh.

Rural Areas Can Take the Lead in Transforming from Energy Consumers into Energy Prosumers

Rooftop photovoltaic energy is prioritized for self-consumption, with excess power exported through the grid: The traditional PV system cannot guide the decarbonization of rural energy use, and rural areas need a new type of power system that allows farmers to enjoy PV power and brings real benefits to them. In this context, the new rural PEDF power system emerges for the so-called PEDF power system, "P" mainly refers to the power supply by renewable energy dominated by RRB rooftop PV; "E" refers to multiple energy storage systems, including heat storage equipment, electric vehicles, electric agricultural machinery, and other electrochemical energy storage equipment; "D" refers to the use of a DC system, which, for one thing, is to reduce losses in power transmission and conversion, and for another, is to facilitate the adoption of low-cost and controller-free control based on busbar voltage signals; "F" refers to the flexible operation of the power system, which is the ultimate purpose. The power system may change its operating state according to user requirements or grid dispatching signals to adjust its power generation and energy consumption curves. Through the rural PEDF system, rooftop PV power is prioritized to meet rural energy demand, with excess power responding to grid scheduling. PEDF reduces the pressure on rural transformer capacity expansion and promotes the full electrification of rural energy use. At the same time, the electrification of agricultural machinery will provide a low-cost method of energy storage for new rural power systems.

Biomass resources are processed into commercialized zero-carbon fuels for economic gain in rural area: Unlike electricity, biomass has low storage and transportation costs. In the context of full rural electrification, biomass resources should be processed into zero-carbon commodity fuels, transforming biomass energy development and utilization from passive consumption to active utilization. Biomass can replace various fossil fuels and is expected to become a main commodity in the future fuel market. Building an economically sustainable "collection, storage, transportation and utilization" integrated industry chain, fully realizing the commercialization of biomass energy, to increase rural income and contribute to rural revitalization.

Key Technologies for Low-Carbon Development in Rural Energy Include PV System Installation, Microgrid Control, PV Heating, Isolated Grid Systems, and Biomass Gasification

Key technologies for low-carbon development in rural energy encompass various aspects, such as PV system installation, microgrid control, PV heating, isolated grid systems, and biomass gasification. Firstly, the installation of rooftop PV systems requires consideration of roof load capacity, tilt angle selection, installation direction, and component spacing, with a focus on post-installation maintenance. Microgrid control technology emphasizes decentralized adaptive control, adjusting equipment operation status through DC voltage signals to ensure stable system operation. PV heating technology, targeting northern rural areas, addresses the mismatch between PV power fluctuations and room heat load through methods like wall heat storage and PV-driven electric heating elements. Isolated grid systems, integrating PV, energy storage, DC and flexibility technology, provide clean and reliable power supply to remote rural areas, suitable for islands, border towns, and nomadic communities with no access to electricity. Additionally, biomass gasification technology converts biomass into combustible gases under high-temperature conditions, offering a green and environmentally friendly path for rural energy utilization. The centralized production and distributed supply integration technology of straw-based natural gas enhance gas production efficiency through rapid chemical pretreatment and multi-material co-fermentation, solving rural energy supply challenges. The application of these technologies will effectively promote the transformation of rural energy structures, achieving low-carbon development goals.

Contents

1	**China's Building Energy Use and GHG Emissions**	1
	1.1 Basic Situation of China's Building Sector	1
	1.1.1 Urban and Rural Demographic	1
	1.1.2 Building Stock	2
	1.2 Demarcation of China's Energy Consumption and GHG Emissions in the Building Sector	4
	1.2.1 Calculation Method of Energy Consumption in the Building Sector	4
	1.2.2 Calculation Method of Carbon Emissions in the Building Sector	7
	1.3 Energy Consumption of China's Building Sector	8
	1.3.1 Building Operation Energy Consumption	8
	1.3.2 Embodied Energy Consumption of Building Sector	16
	1.4 GHG Emissions of China's Building Sector	18
	1.4.1 Carbon Dioxide Emissions from Building Operation	18
	1.4.2 Embodied CO_2 Emission of Building Sector	22
2	**Comparison of Energy Consumption and Carbon Emissions from Building Operation Between China and Other Countries**	27
	2.1 Energy Consumption and GHG Emissions in the Global Building Sector	27
	2.1.1 Energy Consumption by Global Building Operation	27
	2.1.2 Boundaries and Comparative Study Methods of Building Energy Consumption and Emissions	29
	2.2 Energy Consumption and Carbon Emissions in the Building Sector of Different Countries	31
	2.2.1 Building Operation Energy Consumption of Different Countries	31
	2.2.2 Carbon Emissions from Building Operation in Different Countries	33

3	**Current Energy Use Situation and Energy-Saving Paths of Rural Buildings**		37
	3.1	Definitions of Concepts Related to Rural Area	37
	3.2	Current Situation of Rural Building Energy Consumption	39
	3.3	Clean Winter Heating Campaign in Northern China and Its Effects	50
		3.3.1 Energy Conservation and Emission Reduction Effects	52
		3.3.2 Environmental and Health Benefits	63
		3.3.3 Analysis of Clean Heating Problems in Rural Areas	67
	3.4	Analysis of Heating Modes and Energy-Saving Paths of Northern RRBs	70
		3.4.1 "Partial-Time, Partial-Space" Mode is a Common Characteristic of RRB Heating	71
		3.4.2 Rational Thermal Insulation of the Building Envelope of RRBs is the Fundament of Building Energy Efficiency	75
		3.4.3 Flexible Regulation of Heating Equipment is the Core of Building Energy Efficiency	85
	3.5	Summary	92
4	**Key Solution to China's Energy Issue: Energy Revolution Initialized in Rural Areas**		95
	4.1	Transition to Zero-Carbon in China's Electricity Sector	96
	4.2	Alternatives to Fossil Fuels and Supply of Zero-Carbon Fuels	98
	4.3	Rural Areas to Play a Dominant Role in the Future Energy Structure System	99
	4.4	Contribution of Initiating Energy Revolution in Rural Areas to Carbon Neutrality in China	102
	4.5	As the Breakthrough Point of China's Energy Revolution, Rural Areas are Set to Lead the Way	105
	4.6	Practical Significance of Rural Energy Revolution	107
	4.7	Summary	111
5	**Potential and Characteristics of Rural Distributed Zero-Carbon Energy in China**		113
	5.1	Definition of Rural Distributed Zero-Carbon Energy	113
	5.2	Analysis of Total Quantity and Distribution of Rural Rooftop PV Resources in China	114
		5.2.1 Important Value of the Rural Rooftop PV Resources	114
		5.2.2 Method for Analyzing the Power Generation Potential of Microscopic Rooftop PV System	115
		5.2.3 Analysis of the Power Generation Potential of Macroscopic Rooftop PV System	122

	5.3	Analysis of the Biomass Resources of the Rural Areas in China and Their Distribution	130
		5.3.1 Important Value of the Rural Biomass Resources	130
		5.3.2 Method for Analyzing Spatial Distribution of Microscopic Biomass Energy Resources	132
		5.3.3 Analysis of the Potential of Macroscopic Biomass Resources	140
	5.4	Total Quantity and Development Potential of Rural Hydropower Resources in China	145
		5.4.1 Overview of Rural Hydropower	145
		5.4.2 Green Transformation of Rural Hydropower	148
		5.4.3 Development Potential of Rural Hydropower in China	149
	5.5	Summary	150
6	**How Rural Areas Can Take the Lead in Transforming from Energy Consumers into Energy Prosumers**		153
	6.1	New Rural PEDF Power System	154
		6.1.1 Technical Characteristics of New Rural PEDF Power System	156
		6.1.2 Economic Analysis of New PEDF Power System	159
		6.1.3 Analysis of Popularization Mode of PEDF New Power System	160
		6.1.4 Impact of New Rural PEDF Power System on Energy System Structure	162
		6.1.5 Policy Suggestions for the New PEDF Power System	164
	6.2	Electrification Solutions for Rural Energy Use in the Future	166
		6.2.1 Current Situation of Electrification of Energy Use in Rural Areas	167
		6.2.2 Electrification in Heating	170
		6.2.3 Evaluation on Electrification in Energy Consumption for Living in Rural Areas—Taking Cooking Energy as an Example	171
		6.2.4 Carbon Emissions Related to Agricultural Machinery and Evaluation on Electrification of Agricultural Machinery	174
		6.2.5 Suggestions for Increasing the Level of Electrification of Energy Use in Agriculture and Rural Areas	184
	6.3	Rural PEDF Technical Points and Presentation of Village-Level Demonstration Cases	185
		6.3.1 Basic Information	186
		6.3.2 RRB Heating Solution	186
		6.3.3 RRB Cooking Solution	188

		6.3.4	Household Microgrid Solution	189
		6.3.5	Village-Level Microgrid Design Scheme	193
		6.3.6	Production Load and Energy Storage Solutions	196
		6.3.7	Summary of Technical Schemes in the Demonstration Village	198
	6.4	Ways of Commercializing Rural Biomass Fuel		199
		6.4.1	Consumption of Rural Biomass	199
		6.4.2	Biomass Fuel Development Methods	207
		6.4.3	Biomass Energy "Production-Supply-Sales-Service" Trading System	210
7	**Key Technologies for Low-Carbon Development of Rural Energy**			**213**
	7.1	Rural House Rooftop PV Installation Technology		213
	7.2	Rural Microgrid Control Technology		216
	7.3	PV Heating Technology for Rural Areas		218
	7.4	Isolated Grid PEDF System Technology		223
		7.4.1	Introduction of Isolated Grid PEDF System	223
		7.4.2	Isolated Grid PEDF System Application	224
	7.5	PEDF Power Storage-Swapping-Testing Integration Technology for Rural Areas		227
		7.5.1	Proposing of the PEDF Storage-Swapping-Testing Integration Technology	227
		7.5.2	Technical Principles of PEDF Storage-Swapping-Testing Integration System for Rural Areas	228
		7.5.3	Technical Characteristics of PEDF Storage-Swapping-Testing Integration System for Rural Areas	229
		7.5.4	Application of PEDF Charging-Swapping-Testing Integration Technology for Rural Areas	230
		7.5.5	Pilot Application and Development of Rural PV-Storage-Swapping-Testing Integration Technology	231
	7.6	Tracking-Free Solar Heat Collection Technology to Meet the Seasonal Heating Requirements for RRB		234
		7.6.1	Technical Limitations of Traditional Solar Water Heating Systems	234
		7.6.2	Principle of Heat Collection Technology Based on Seasonal Adaptability of Heating Requirements	235
		7.6.3	Technical Assembly Modes in Different Application Scenarios	236
		7.6.4	Plant Performance Tests	237
	7.7	Summary		239

7.8	Biomass Pyrolysis and Gasification Technology		240
	7.8.1	Basic Principles of Pyrolysis and Gasification Technology	240
	7.8.2	Product Performance	244
7.9	Integration of Natural Gas Production from Straws and Distributed Gas Supply		246
	7.9.1	Technical Principle	246
	7.9.2	Technical Characteristics	250

References 253

List of Figures

Fig. 1.1	Population growth in China by year (2001–2022)	2
Fig. 1.2	Newly built and demolished building stock of civil buildings in China (2007–2022)	3
Fig. 1.3	Newly built building stock of P&C buildings according to different functions (2001, 2022)	3
Fig. 1.4	China's existing building stock (2001–2022). *Source* Estimation results from CBEEM of the Building Energy Research Center (BERC), Tsinghua University. *Note* The newly built building stock entered in the model is the data under the statistical standards for construction enterprises as specified in the China Statistical Yearbook on Construction	4
Fig. 1.5	Primary energy consumption and total electricity consumption of building operation in China (2010–2022)	9
Fig. 1.6	Building operation energy consumption in China (2022). *Note* Electricity, heat, and fuels are uniformly converted into primary energy, which is measured in standard coal, and the electricity consumption is converted into primary energy consumption calculated in standard coal based on the annual average coal consumption for power supply in China. The conversion coefficient in 2022 was 301 gce/kWh	10
Fig. 1.7	Changes in total building energy use and energy use intensity by year (2002–2022)	11
Fig. 1.8	Embodied energy use of China's civil buildings (2004–2022). *Source* Estimation by BERC, Tsinghua University. This figure only covers civil building construction	16

Fig. 1.9	Embodied energy consumption of China's construction sector (2004–2022). *Source* Estimation by BERC, Tsinghua University. *Note* The construction sector involves the construction of civil buildings, production buildings, and infrastructures	17
Fig. 1.10	Energy consumption for building material production	18
Fig. 1.11	Carbon dioxide emissions from building operation (2022)	19
Fig. 1.12	Carbon dioxide emissions from building operation in China (2022)	22
Fig. 1.13	Carbon emissions of building energy use (2012–2022)	23
Fig. 1.14	Embodied carbon emissions of civil buildings in China (2004–2022). *Source* Estimation by BERC, Tsinghua University. *Note* Civil building construction is included only	24
Fig. 1.15	Embodied carbon emissions of the construction sector in China (2004–2022). *Source* Estimation by BERC, Tsinghua University. *Note* The construction sector involves the construction of civil buildings, production buildings, and infrastructures	25
Fig. 2.1	Energy use and CO_2 emissions in the global building sector (2021). *Source* 2022 Global Status Report for Buildings and Construction, International Energy Agency. *Note* The construction sector involves civil building construction, production building construction, and infrastructure construction. This figure uses the terminal energy consumption data provided by IEA, which is obtained through the direct addition of heat consumption for heating, electricity consumption of buildings, and terminal use of various energy types. The electricity consumption is converted to primary energy using a calorific value equivalent method. This conversion method is different from that used in the comparison of building energy consumption among countries in the ensuing part of this research, so it should be treated differently in data comparison	28
Fig. 2.2	China's building energy use and CO_2 emissions (2022). *Source* Estimation with CBEEM of BERC, Tsinghua University. *Note* The construction sector involves the construction of civil buildings, production buildings, and infrastructures. The diagram on the right illustrates the structure of China's total social carbon emissions (including energy-related and industrial process emissions)	29

Fig. 2.3	Comparison of building operation energy consumption in different countries (electricity equivalent method). *Source* CBEEM of BERC, Tsinghua University; World Energy Balances, Energy Efficiency Indicators database (2023 edition) of IEA; WDI database of the World Bank; Satish Kumar (2019) of India. Data from 2020 for Canada, and data from 2020 for other countries	32
Fig. 2.4	The electrification rate of energy use in the construction sector by country (2001–2021)	33
Fig. 2.5	Per capita carbon emissions comparison by country (2021)	33
Fig. 2.6	Comparison of per capita carbon emissions of different countries (2021). *Source* Data of countries in 2021 as provided in the CO_2 Emissions from the Fuel Combustion Highlights 2023 database, IEA. Data from China are the results of CBEEM of BERC, Tsinghua University	34
Fig. 2.7	Comparison of the trends in carbon emissions from building operations between china and foreign countries (2001 and 2021). *Source* Data of countries in 2021 as provided in the CO_2 Emissions from the Fuel Combustion Highlights 2023 database, IEA. Data from China are the results of CBEEM of BERC, Tsinghua University	35
Fig. 3.1	Distribution of proportions of different energy consumption of rural buildings in some provinces (autonomous regions and municipalities directly under the Central Government) of China. **a** Northern region; **b** Southern region	44
Fig. 3.2	Per capita annual physical consumption of bulk coal in rural buildings in different regions of China (I). **a** "Beijing-Tianjin-Hebei" and surrounding areas; **b** Other northern provinces except for "Beijing-Tianjin-Hebei" and surrounding areas. Per capita annual physical consumption of bulk coal in rural buildings in different regions of China (II). **c** Provinces in the Yangtze River Basin; **d** Other southern provinces except for those in the Yangtze River Basin	46
Fig. 3.3	Typical rural houses in a village in Inner Mongolia	49
Fig. 3.4	Sunrooms and double-window rural houses in a village in Inner Mongolia	49
Fig. 3.5	Winter heating facilities for farmer households in a village	49
Fig. 3.6	Percentage contribution of $PM_{2.5}$ emission reduction caused by the change in domestic energy use in rural areas of different regions in China from 2014 to 2022	64
Fig. 3.7	Changes in the mean annual concentration of $PM_{2.5}$ in "Beijing-Tianjin-Hebei" and surrounding areas in the past 10 years	65

Fig. 3.8	Instrument used for monitoring in a village in Fangshan District, Beijing in 2013 and 2018	66
Fig. 3.9	Daily mean concentrations of $PM_{2.5}$ in the air from January to March in a village in Fangshan District, Beijing in 2013 and 2018	66
Fig. 3.10	Comparison of floor plans of common urban and rural residential buildings (m). **a** Urban residential building; **b** Rural residential building	72
Fig. 3.11	Measured actual heating needs of different rooms of a rural household in Beijing. **a** Typical floor plan; **b** onsite test diagram; **c** time-dependent operation rates of the air-to-air heat pumps in the rooms; **d** heating utilization rates of the rooms; **e** power of the air-to-air heat pumps in the main heated rooms on typical days and the interior temperature curve	75
Fig. 3.12	Flow diagram for menu-based energy efficiency retrofitting of building envelope	77
Fig. 3.13	Comparison of the results of menu-based energy efficiency retrofitting solutions	78
Fig. 3.14	Plan of an RRB in Shanxi	79
Fig. 3.15	Plan of an RRB in Heilongjiang	83
Fig. 3.16	Makeshift sunroom of an RRB in Northeast China	84
Fig. 3.17	Comparison between makeshift sunroom temperature and outdoor temperature	84
Fig. 3.18	Heat map of 3-year heating demand distribution of air-to-air heat pumps in each room of an actual RRB in northern China. **a** East bedroom; **b** Middle bedroom; **c** Bathroom; **d** West bedroom; **e** Kitchen and dining room; **f** Living room	87
Fig. 3.19	Floor plan of RRB for comparison testing of heating equipment **a** Floor plan; **b** RRB photo	89
Fig. 3.20	Two heating systems and test instruments. **a** Air-to-water heat pump; **b** Control panel of air-to-water heat pump; **c** Power test module of air-to-water heat pump; **d** Indoor unit of the air-to-air heat pump in the master bedroom; **e** Outdoor unit of the air-to-air heat pump; **f** Test power meter of the air-to-air heat pump	90
Fig. 3.21	Running power of the air-to-air heat pump in each room*. **a** Guest bedroom 1; **b** Living room; **c** Master bedroom; **d** Guest bedroom 2. *Note* The color version of this figure can be viewed by scanning the QR code in the Table of Contents	90
Fig. 3.22	Running power of the air-to-water heat pump	91

Fig. 3.23	Heating diagram of air-to-air heat pump, hourly heating probability, and room temperature (*Note* Blue represents the heating period)*. **a** Guest bedroom 2; **b** Master bedroom; **c** Living room; **d** Guest bedroom 1	91
Fig. 3.24	Daily room temperature in each room during use of air-to-water heat pump*. **a** Guest bedroom 2; **b** Master bedroom; **c** Living room; **d** Guest bedroom 1	91
Fig. 4.1	Rural energy demand and carbon emission prediction under carbon–neutral scenario*. **a** Energy demand; **b** Carbon emissions from energy	104
Fig. 4.2	Rural energy demand and carbon emission prediction under base scenario*. **a** Energy demand; **b** Carbon emissions from energy	104
Fig. 5.1	Typical distributed zero-carbon energy sources in rural areas. **a** Rural rooftop PV; **b** biomass; **c** rural small hydropower stations	114
Fig. 5.2	Schematic diagram for typical rural building	116
Fig. 5.3	Two types of open-source remote sensing satellite images and the actual labels for the corresponding roofs	116
Fig. 5.4	Structural diagram for corrected U-Net model	117
Fig. 5.5	Processing of identifying roofs. U-roof perimeter; S-roof area; A-equivalent length of roof; B-equivalent width of roof	118
Fig. 5.6	Schematic diagram for two installation methods for rooftop PV panel on a flat roof. **a** Optimum tilt angle; **b** parallel roof installation	118
Fig. 5.7	Comparison between identification result and measurement result	119
Fig. 5.8	Identification result of the corrected U-Net model	120
Fig. 5.9	Calculated rooftop PV power generation potential of village A	121
Fig. 5.10	Calculated rooftop PV potential of the villages in the town territory	123
Fig. 5.11	Study process for nationwide rural rooftop PV resource potential	123
Fig. 5.12	Satellite image and country label of typical villages	124
Fig. 5.13	Comparison of different provinces in terms of village contour area	125
Fig. 5.14	Contour identification result of typical villages and rural roof identification result. **a** Northern village; **b** southern village $R_c = S_p/S_D(4-1)$	126
Fig. 5.15	Visible light based part of the remote sensing image of various surface features	132
Fig. 5.16	Average spectral brightness values of various surface features sampled **a** Remote sensing image of March; **b** Remote sensing image of August	133

Fig. 5.17	*NDVI* value distribution diagram corresponding to the remote sensing image of August in the study area	134
Fig. 5.18	Schematic diagram for growth cycle of summer corn and winter wheat	135
Fig. 5.19	Satellite image and NDVI image of winter wheat at main time nodes	136
Fig. 5.20	Satellite image and NDVI image of summer corn at main time noes	136
Fig. 5.21	Variation pattern of *NDVI* statistics of the crops in different time phases in the sampling areas. **a** Summer corn; **b** winter wheat	137
Fig. 5.22	Variation pattern of the *NDVI* statistics of the other surface features in the sampling areas. **a** Other vegetation; **b** structures; **c** water body	138
Fig. 5.23	Decision tree model used for extracting the crops in the study area	138
Fig. 5.24	Distribution of the crops identified by the method proposed	139
Fig. 5.25	Winter wheat extraction results of some areas	139
Fig. 5.26	Number and total installed capacity of the rural hydropower stations in China in 2011–2020	146
Fig. 6.1	Installed capacity of residential DG PV in some provinces (autonomous regions and municipalities directly under the Central Government) (as of the first half of 2023). *Data Source* Official website of the National Energy Administration of China	155
Fig. 6.2	Comparison of two types of PV power systems	156
Fig. 6.3	Topological graph of the new rural PEDF power system	157
Fig. 6.4	Schematic diagram of the new rural PEDF power system	159
Fig. 6.5	Analysis of the roles of the participants in rural PEDF new power system	161
Fig. 6.6	Energy structures in some regions in China in 2022*. *Note* The data is from the official website of the National Energy Administration	163
Fig. 6.7	Comparison of rural rooftop PV power generation potential and thermal energy production in various regions	164
Fig. 6.8	Electrification rates of agriculture and rural residents' life in different regions of China in 2020. *Note* Relevant statistical data for Xizang, Hong Kong, Macao, and Taiwan are missing	167
Fig. 6.9	Changes in the electrification rates of agriculture and rural residents' life in China from 2016 to 2020	168
Fig. 6.10	Electrification rates of agriculture and rural residents' life in the four major regions of China from 2018 to 2020	168
Fig. 6.11	Quantity of household appliances owned by every 100 rural households	170

Fig. 6.12	Current situation of rural cooking in different regions of China	173
Fig. 6.13	Changes in carbon emissions from agricultural machinery in China from 2000 to 2020	175
Fig. 6.14	Distribution of cumulative carbon emissions from agricultural machinery from 2000 to 2020	175
Fig. 6.15	Power prediction and electrification potential evaluation of agricultural machinery. **a** Prediction of total power of agricultural machinery; **b** Penetration rate of electric agricultural machinery; **c** Power for the electrification of agricultural machinery	179
Fig. 6.16	Carbon emission reduction benefits of the electrification of agricultural machinery under different scenarios	180
Fig. 6.17	Classification of agricultural machinery	180
Fig. 6.18	Power of common agricultural machinery	181
Fig. 6.19	Levels of operation of different types of agricultural machinery	182
Fig. 6.20	Estimation of energy storage potential of agricultural machinery. **a** Agricultural machinery; **b** Specific equipment	182
Fig. 6.21	Three different power system topologies	189
Fig. 6.22	No PV power scenario	192
Fig. 6.23	Low light scenario	192
Fig. 6.24	High light scenario	193
Fig. 6.25	Operating state of village-level microgrid during the clear day	194
Fig. 6.26	Operating state of village-level microgrid on consecutive clear nights	195
Fig. 6.27	Operating state of village-level microgrid on rainy days	195
Fig. 6.28	Comparison of village-level microgrid operation under different energy storage capacities	196
Fig. 6.29	Standardized battery swap cabinet	197
Fig. 6.30	Topology of the new PEDF power system in Village A	199
Fig. 6.31	"Five methods" of straw utilization in China and open burning of straws. **a** Utilization of straws as fuel; **b** Utilization of straws as manure; **c** Utilization of straws as feed; **d** Utilization of straws as raw materials; **e** Utilization of straws as base materials; **f** Open burning of straws	200
Fig. 6.32	Comparison of fertilizer application amount per unit area of farmland	204
Fig. 7.1	Rooftop PV installation components	214
Fig. 7.2	DC busbar voltage control principle	216
Fig. 7.3	Mismatch between PV power and rural house heating load	219
Fig. 7.4	Schematic diagram of PV heating system based on wall heat storage	220

Fig. 7.5	Heating test effect*	221
Fig. 7.6	PV-driven air-to-air heat pump system **a** Air-to-air heat pump indoor unit; **b** Control system cabinet	222
Fig. 7.7	PV station in Holy Elephant Gate, Nam Co	225
Fig. 7.8	Zhuhai Hebao Island 5G base station	226
Fig. 7.9	Off-grid villa demonstration project in Nigeria	226
Fig. 7.10	Architecture of PEDF storage-swapping-testing integration system for rural areas	228
Fig. 7.11	Application scenario of low-speed electric vehicles in rural areas	230
Fig. 7.12	Application scenario of integrated PV-storage-charging-swapping-testing cabinet in the transformer coverage area	231
Fig. 7.13	PV future roof DC community DC 48 V storage-swapping research pilot project	232
Fig. 7.14	Rural electrification pilot project of State Grid—Chongqing Baihui Park	233
Fig. 7.15	System problems caused by excessive heat in summer	234
Fig. 7.16	Characteristic difference of solar radiation in winter and summer [A case study of Beijing (37.45°N)]	235
Fig. 7.17	Schematic diagram of an angle-constrained concentrator	236
Fig. 7.18	Schematic diagram of the new-type collector	237
Fig. 7.19	Assembly modes in different application scenarios	237
Fig. 7.20	Comparison of optical efficiency tests in winter and summer	238
Fig. 7.21	Transverse comparison of performance between new-type collectors and traditional collectors	239
Fig. 7.22	Application effects of new-type collectors vs. traditional flat plate collectors in different cities. **a** Collectors in winter; **b** Collectors in summer	240
Fig. 7.23	Schematic diagram of biomass gasification principle	241
Fig. 7.24	Technical route diagram for biogas-based natural gas production from straws	247
Fig. 7.25	Process flow diagram for pressurized water washing	249
Fig. 7.26	Granular organic compound fertilizer production system	250

List of Tables

Table 1.1	Building operation energy consumption in China in 2021	9
Table 2.1	Energy conversion factor benchmark in the calculation of total energy consumption of buildings in various countries	30
Table 3.1	Consumption of different types of energy in rural household energy consumption in some provinces (autonomous regions and municipalities directly under the Central Government) of China in 2022	41
Table 3.2	Pilot cities and projects for clean heating in northern China during 2017–2022	53
Table 3.3	Comparison of total consumption of different types of energy in the energy use of rural buildings in some provinces (autonomous regions and municipalities directly under the Central Government) of China in 2022 and 2014	56
Table 3.4	Emission factors of $PM_{2.5}$ and other gas pollutants from different stoves	62
Table 3.5	Information on an RRB in Shanxi	80
Table 3.6	Energy efficiency retrofitting solutions for an RRB in Shanxi (continuous heating)	81
Table 3.7	Energy efficiency retrofitting solutions for an RRB in Shanxi (intermittent heating)	82
Table 3.8	Information on an RRB in Heilongjiang	83
Table 3.9	Energy efficiency retrofitting solution for an RRB in Heilongjiang	85
Table 3.10	Room size and daily average heating duration of an actual RRB in northern China	88
Table 4.1	National total power consumption and future forecast (unit: PWh)	96
Table 4.2	China's power supply structure and annual energy production forecast	97

Table 4.3	Predetermined values of key factors influencing rural energy consumption and production under different development scenarios	103
Table 5.1	Annual total PV power generation potential of the villages and households in the study area	122
Table 5.2	Calculated proportion coefficient of rural buildings and roof area in some of the provinces (autonomous regions or municipalities) in China	127
Table 5.3	Potential rooftop PV installation area, energy production, and installed power in the rural areas in some of the provinces (autonomous regions or municipalities) in China	129
Table 5.4	Comparison of the identified area and the actual planting area of the towns in the study area	140
Table 5.5	Harvestable quantity of different types of crop straws in some of the provinces (autonomous regions or municipalities) (in 10,000 t)	142
Table 5.6	Proportion and production output of residual materials generated by processing of various crops in China	144
Table 5.7	Summary of rural biomass resources in China	144
Table 5.8	Installed capacity and annual energy production of the hydropower projects in the regions [24]	147
Table 6.1	Comparison of household power topology schemes	190
Table 6.2	Advantages and disadvantages of different consumption methods of straws	201
Table 6.3	Emission factors of open burning of straws	202
Table 6.4	Maximum consumption and current consumption of different consumption methods of straws	203
Table 6.5	CH_4 and N_2O emission factors for different biomass consumption methods	206
Table 6.6	Gas composition and calorific value of different raw materials	209
Table 7.1	Coordinated control strategy for converters under different operating modes	218
Table 7.2	Comparison of winter and summer data in different cities	241
Table 7.3	The main composition of gas and lower calorific value	244
Table 7.4	Gaseous compositions and calorific values of different raw materials	245
Table 7.5	Parameters of rice husk and wood chip	245
Table 7.6	Properties of the fluids extracted from rice husk	246

Chapter 1
China's Building Energy Use and GHG Emissions

1.1 Basic Situation of China's Building Sector

1.1.1 Urban and Rural Demographic

In recent years, urbanization has grown rapidly in China. In 2022, China had an urban population of 921 million and a rural population of 491 million, with the urbanization rate rising from 37.7% in 2001 to 65.2%, as shown in Fig. 1.1.

China's urbanization rate has been increasing year by year but the rate of growth has gradually slowed down, and the direction of population mobility has shifted from urban–rural migration in the immediate vicinity to inter-urban migration on a larger scale. According to relevant studies, the overall pattern of population movement in China's provincial areas continues to show a flow from the west to the east and from the north to the south, but in recent years there has been a phenomenon of population return to the central and western provinces.[1] According to the China Statistical Yearbook, in 2022, provinces and cities such as Guangdong, Beijing and Shanghai experienced negative growth in their resident populations, while more cities in central and western provinces, such as Jiangxi and Hubei, experienced significant growth. Changes in population flows and economic growth in central and western provinces and their capital cities in recent years reflect China's regional economies moving toward greater coordination.

The development pattern of cities and towns has also gradually shifted from incremental expansion to quality enhancement along with population movement and urban development. According to a study conducted by the Institute of Spatial Planning & Regional Economy, National Development and Reform Commission P.R.C, for a typical country, after the urbanization rate reaches 60 percent, the overall framework of the city is basically established, the main infrastructure is basically completed, and

---

[1] Zhikai Wang. New trends in China's population mobility and the coordinated development of the regional economy.

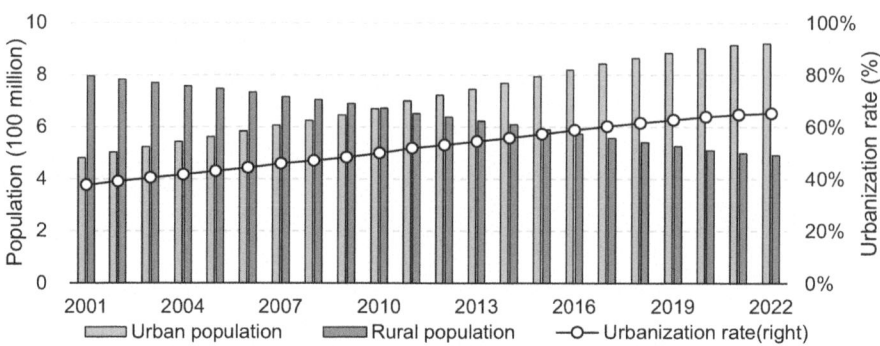

Fig. 1.1 Population growth in China by year (2001–2022)

the mode of urban development shifts to a mode in which "incremental expansion and renewal of stock are given equal importance".[2] In August 2021, the Circular of the Ministry of Housing and Urban–Rural Development of the People's Republic of China on Preventing the Problem of Large-scale Demolition and Construction in the Implementation of Urban Renewal Actions (Jianke 2021 No. 63) emphasized that urban renewal should change the mode of urban development and construction, adhere to the concomitant measures of "retaining, changing and demolishing", and focus on retaining and utilizing the enhancement of the main, to strengthen repairs and renovation, to make up for the shortcomings of the city, and focus on the enhancement of functionality, so as to strengthen the vitality of the city. At the same time, the phenomenon of large-scale demolition and construction is strictly controlled; in principle, the demolition of building area in an urban renewal unit (area) or project should not be greater than 20% of the total building area of the status quo; in principle, the ratio of demolition and construction in an urban renewal unit (area) or project should not be greater than 2. From the point of view of the sustainable development of urbanization, the question of how cities can achieve their own renewal while avoiding major demolition and construction should be given sufficient attention.

1.1.2 Building Stock

Rapid urbanization drives continuous development of the construction sector, and the scale of China's construction sector has been expanding. From 2007 to 2022, thanks to the rapid growth of building construction in China, the floor space greatly expanded in urban and rural areas. Specifically, from 2007 to 2014, China's newly built building stock of civil buildings grew rapidly, while it has been declining slowly each year from 2014 to the present. The newly built building stock of urban residential

---

[2] Jiyuan Wang. Characteristics of real estate transformation trends in typical countries after reaching 60% urbanization rate.

1.1 Basic Situation of China's Building Sector

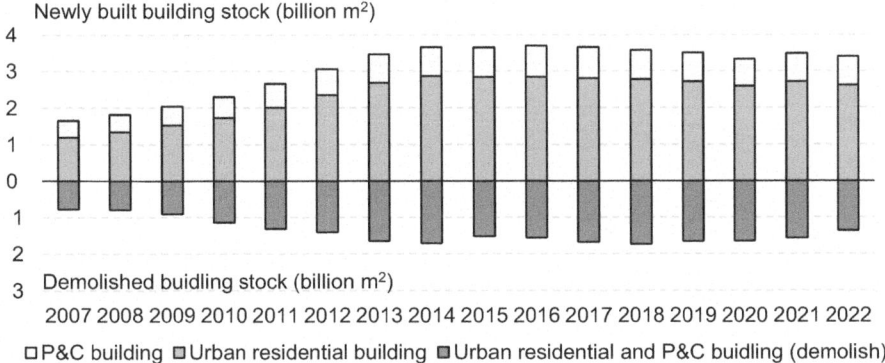

Fig. 1.2 Newly built and demolished building stock of civil buildings in China (2007–2022)

building and P&C building slowly declined from around 3.6 billion m^2 in 2014 to 3.4 billion m^2 in 2022 (Fig. 1.2). Driven by the change of the number of construction projects, the demolished building stock of urban residential areas and P&C buildings increased rapidly from 700 million m^2 in 2007 to approximately 1.6 billion m^2 in 2018, and then declined slowly to 1.4 billion m^2 in 2022.

In 2022, residential buildings and non-residential buildings made up about 79 and 21% of the newly built building stock of civil buildings in China respectively. According to the difference in building functions, public and commercial (P&C) buildings can be categorized into offices, hotels, malls, hospitals, schools, and others. From 2001 to 2022, the main types completed each year were dominated by offices, malls, and schools. In 2022, the total newly built building stock of these three types accounted for about 68% of that of public and commercial (P&C) buildings, with the proportions of malls, office buildings, and schools at 30, 18, and 20% respectively. For the remaining types, hospitals and hotels had a smaller proportion of 6 and 3% respectively (Fig. 1.3).

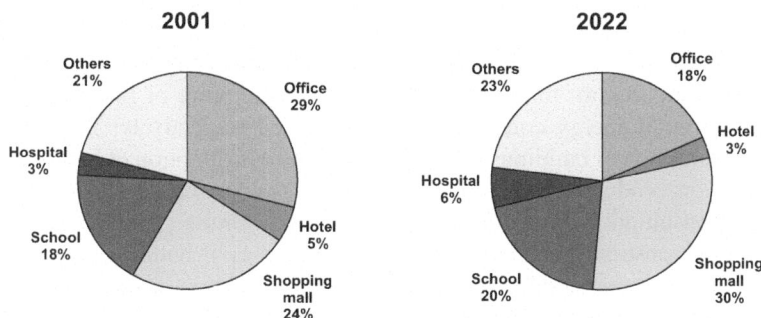

Fig. 1.3 Newly built building stock of P&C buildings according to different functions (2001, 2022)

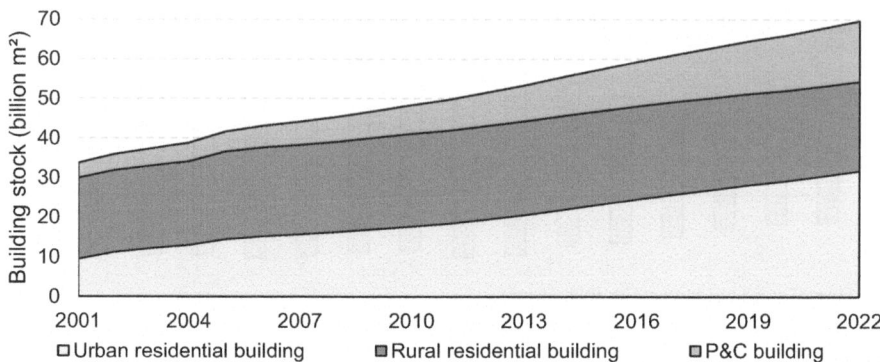

Fig. 1.4 China's existing building stock (2001–2022). *Source* Estimation results from CBEEM of the Building Energy Research Center (BERC), Tsinghua University. *Note* The newly built building stock entered in the model is the data under the statistical standards for construction enterprises as specified in the China Statistical Yearbook on Construction

Large-scale building construction activity has led to the rapid growth of China's building stock every year. In 2022, the total building stock in China was about 69.6billion m^2, including urban residential buildings accounted for 31.8 billion m^2, rural residential buildings accounted for 22.4 billion m^2, and P&C buildings accounted for 15.4 billion m^2 (Fig. 1.4). The floor area for northern urban heating stood at 16.7 billion m^2.

1.2 Demarcation of China's Energy Consumption and GHG Emissions in the Building Sector

1.2.1 Calculation Method of Energy Consumption in the Building Sector

Energy consumption for the building sector covers different phases of buildings life cycle. In this report, the embodied energy consumption of buildings and the building operation energy consumption are analyzed respectively. The embodied energy consumption of buildings refers to the energy consumption of building material exploration, production, transportation and on-site construction and also includes the energy consumption during building demolition. In China's statistical standards, civil building construction, production building (non-civil building) construction, and infrastructure construction are included in the construction sector, so their energy consumption is collectively known as the embodied energy consumption related to the construction sector. Based on the China Building Energy and Emission Model of BERC, Tsinghua University, this book provides the analysis data of the standard for embodied energy consumption of China's construction sector and the standard

1.2 Demarcation of China's Energy Consumption and GHG Emissions ...

for embodied energy consumption of China's civil buildings (see Sect. 1.3.2 for details). Building operation energy consumption, which is the focus of this book, refers to the energy consumption from the operation of civil buildings, including the energy consumed by the provision of heating, ventilation, air conditioning, lighting, cooking, and domestic hot water (DHW) to occupants or users in residential buildings, office buildings, schools, malls, hotels, transportation hubs, recreational and sports facilities, and other non-industrial buildings, and energy consumed by service functions of such buildings. It is very difficult to distinguish the operational energy consumption of buildings that fully serve industrial production processes from the industrial production energy consumption, because the energy consumption in ventilation, air conditioning, and purification of factory buildings including metallurgical factory buildings and integrated circuit or pharmaceutical production factory buildings accounts for a very large proportion in the production energy consumption. However, it is very difficult to include such energy consumption in building energy consumption. Hence, this report does not discuss the buildings designed for production processes but focuses solely on civil buildings serving humans.

Based on our long-term research on the energy consumption of civil building operation in China and given the difference in heating methods in winter between northern and southern China, the difference in architectural forms and lifestyles between urban and rural areas, and the difference in personnel activities and energy use equipment between residential and P&C buildings, this report divides the building energy use in China into four categories, i.e. northern urban heating (NUH) energy use, urban residential building energy use (excluding NUH energy use), public and commercial building energy use (excluding NUH energy use), and rural residential building energy use, which are defined in detail as follows.

(1) **NUH energy use**

This refers to the energy consumption of heating in winter, including all forms of centralized heating and decentralized heating, in provinces, autonomous regions, and municipalities that adopt centralized heating, including all urban areas of Beijing, Tianjin, Hebei, Shanxi, Inner Mongolia, Liaoning, Jilin, Heilongjiang, Shandong, Henan, Shaanxi, Gansu, Qinghai, Ningxia, and Xinjiang as well as part of Sichuan. Heating is also needed in winter in Tibet, western Sichuan, and part of Guizhou, but should be considered separately as the local energy situations, issues, and characteristics are completely different from those in northern China. The reason for the separate calculation of NUH energy use is that centralized heating has been the main heating method in northern urban areas, including a large number of city-level heating networks and community-level heating networks. Different from other categories of building energy use, in which the calculation is based on the consumption of a single building or a single household, NUH energy use is largely related to the structural form and operation mode of a heating system, and the actual value of energy use is counted and calculated in a unified manner based on the heating system, so NUH energy use is taken as a separate category and treated differently from other categories of building energy use. Based on the form and scale of heat source systems, the current heating systems can be classified into centralized heating systems, which adopt such methods as large- and medium-scale coal-fired combined heat and power (CHP), large- and medium-scale gas-fired CHP, small-scale coal-fired CHP, small-scale gas-fired CHP, large coal-fired boilers, large gas-fired boilers, district coal-fired boilers, district gas-fired boilers, heat pump for centralized heating, residual heat from nuclear power and industrial residual heat, and household heating systems, which adopt such methods as household gas furnaces, household coal furnaces, air conditioners and heat pumps for decentralized heating, and direct electric heating. The main types of energy sources used include coal, gas, and electricity. This report studies the primary energy consumption, including the primary energy or electricity consumption at heat sources and the electricity consumption of equipment (fans and water pumps) serving heating systems. Such energy consumption can also be divided into the conversion loss from heat sources and heating stations, the heat loss and energy consumption of the distribution pipe networks, and the final heat gains of buildings.

(2) **Urban residential building energy use (excluding NUH energy use)**

This refers to the energy consumption of urban residential buildings, except for the heating energy consumption in northern China. In terms of energy end-users, it includes energy consumption of household appliances, air conditioners, lighting, cooking, and domestic hot water, as well as energy consumption of heating in winter in provinces, autonomous regions, and municipalities in the hot-summer and cold-winter (HSCW) zone. The main types of commercial energy sources used in urban residential buildings include electricity, coal, natural gas, liquefied petroleum gas (LPG), and city gas. Decentralized heating is mostly adopted in winter in the HSCW zone, and the energy consumption of the following heating methods all falls into this category: building space heating methods such as air source heat pumps and direct electric heating, and local heating methods such as fire pans, electric blankets, and electrical hand warmers.

(3) **Public and commercial building energy use (excluding NUH energy use)**

The public and commercial (P&C) buildings here refer to buildings where people carry out various public activities including office buildings, commercial buildings, tourism buildings, scientific research, educational, cultural, and medical buildings, communication buildings, and transportation buildings in urban and rural areas. Except for NUH energy consumption, the energy consumption of activities in buildings includes the energy consumption of air conditioning, lighting, sockets, elevators, cooking, and service facilities, as well as the energy consumption of heating of urban P&C buildings in winter in the HSCW zone. The types of commercial energy sources used in P&C buildings include electricity, gas, fuel oil, and coal.

(4) **Rural residential building energy use**

This refers to the energy consumption of rural households, including cooking, heating, cooling, lighting, hot water, household appliances, etc. The main types of energy sources used in rural residential buildings include electricity, coal, LPG, gas, and biomass energy (straw and firewood). The consumption of biomass energy is not included in the national macrostatistics of energy. However, as an important part of rural residential building energy use, it will be listed separately in this report.

In this report, actual consumption of electricity other types of energy is counted and calculated separately whenever possible. If they have to be combined, all energy sources will be converted into primary energy sources for addition, namely, the electricity consumption will be converted into primary energy consumption calculated in standard coal based on the annual average coal consumption for power supply in China. As to the CHP method of centralized heating source for building operation, the input fuels are allocated based on the exergy values of output electricity and heat according to relevant provisions of the *Standard for Energy Consumption of Building* (GB/T 51,161-2016). In this report, the conversion coefficient for the exergy of heat is calculated based on an ambient temperature of 0 °C and a supply/return water temperature of 110/50 °C, and the conversion coefficient for heat is 0.22.

1.2.2 Calculation Method of Carbon Emissions in the Building Sector

The embodied carbon emissions of buildings include the carbon emissions from the building material production, transportation, on-site construction, and demolition of civil buildings. In China's statistical standards, civil building construction, production building (non-civil building) construction, and infrastructure construction are included in the construction sector, so their carbon emissions are collectively known as the embodied carbon emissions related to the construction sector. Based on the China Building Energy and Emission Model of BERC, Tsinghua University, this report provides the analysis data of the standard for embodied carbon emissions of China's construction sector and the standard for embodied carbon emissions of China's civil buildings (see Sect. 1.4.2 for details).

Carbon emission during building operation mainly includes the carbon emission from the direct burning of fossil fuels and the indirect use of non-fossil energy during the operation of buildings, which consists of three main types:

1. **Direct carbon emission**: It refers to the direct emission of carbon dioxide in buildings by burning fossil fuels including coal, fuel oil, and gas. The carbon emission can be calculated based on the types of fuels and their different carbon emission factors.
2. **Indirect carbon emission from electricity use**: It refers to the carbon emission during the generation of electricity transmitted into buildings from the outside. The carbon emission can be calculated through the multiplication of the total external electricity used for buildings by the average carbon emission factor of electricity in the power grid, and the PV power generation and electricity consumption in buildings themselves are not counted.
3. **Indirect carbon emission from heating**: It refers to the indirect carbon emission resulting from centralized heating in northern urban areas. The centralized

heating systems in northern urban areas adopt combined heat and power generation or centralized coal- and gas-fired boilers for the supply of heat. In this regard, carbon dioxide emitted by coal- and gas-fired boilers fall into the indirect carbon emission from building heating, while carbon emissions of combined heat and power generation plants are allocated according to exergy values of output electricity and heat. In this report, the conversion coefficient for the exergy of heat is calculated based on an ambient temperature of 0 °C and a supply/return water temperature of 110/50 °C, and the conversion coefficient for heat is 0.22. That is to say, 22% of the output heat is treated as the equivalent electricity, which shares the total carbon dioxide emitted from power plants with the output electricity.

1.3 Energy Consumption of China's Building Sector

1.3.1 Building Operation Energy Consumption

The building energy consumption data in this chapter comes from the results of research with the China Building Energy and Emission Model (CBEEM) built by BERC, Tsinghua University, and is used to analyze the development of building energy consumption and carbon emissions in China. In 2022, the total commercial energy consumption of building operation was 1.12 gigatonnes of coal equivalent (Gtce), accounting for about 21% of the total energy consumption in China, and the commercial energy consumption and biomass energy consumption of buildings amounted to 1.17 Gtce (biomass energy consumption: about 0.05 Gtce), with details given in Table 1.1.

From 2010 to 2022, the total energy consumption of buildings and electricity consumption increased dramatically, as shown in Fig. 1.5. The COVID-19 pandemic slowed down various social activities, and the growth in electricity consumption of buildings in 2020 was slower than that in 2019. However, as production and life returned to normal in 2021, the electricity consumption of buildings rose greatly. The electricity consumption of buildings in the whole society exceeded 2.3 PWh in 2022.

The scale, intensity, and total quantity of the four categories of building energy use are represented in the four blocks in Fig. 1.6 respectively, in which the horizontal axis represents building floor area, and the vertical axis represents energy use intensity per square meter of building. The total area of the four blocks represents the total energy consumption of buildings. In terms of building floor area, urban residential buildings and rural residential buildings have the largest floor area, while the floor areas of NUH and P&C buildings account for about one-fourth and only one-fifth of the total, respectively. However, from the perspective of energy use intensity, the energy use intensity of P&C buildings and NUH is higher than that of the other two categories. Therefore, in terms of total energy use, each of the four categories accounts for about one-fourth of the total building energy consumption. In recent years, the growth in the scale and average energy use intensity of P&C buildings

1.3 Energy Consumption of China's Building Sector

Table 1.1 Building operation energy consumption in China in 2021

Energy use category	Macro parameter	Electricity (PWh)	Fossil fuel (billion tce)	Commercial energy (billion tce)	Primary energy use intensity
NUH	16.7 billion m^2	0.079	0.193	0.217	13 kgce/m^2
Urban residential building (excluding NUH)	31.8 billion m^2	0.6307	0.101	0.291	787 kgce/household
P&C building (excluding NUH)	15.4 billion m^2	1.2707	0.026	0.408	26.5 kgce/m^2
Rural residential building	22.4 billion m^2	0.3481	0.096	0.201	1,070 kgce/household
Total	1.41 billion people 69.6 billion m^2	2.3285	0.416	1.12	

Note In the table, commercial energy consumption means the energy consumption calculated in standard coal converted from electricity, heat, and fuels, and electricity consumption specifically refers to the electricity consumption in the building energy use

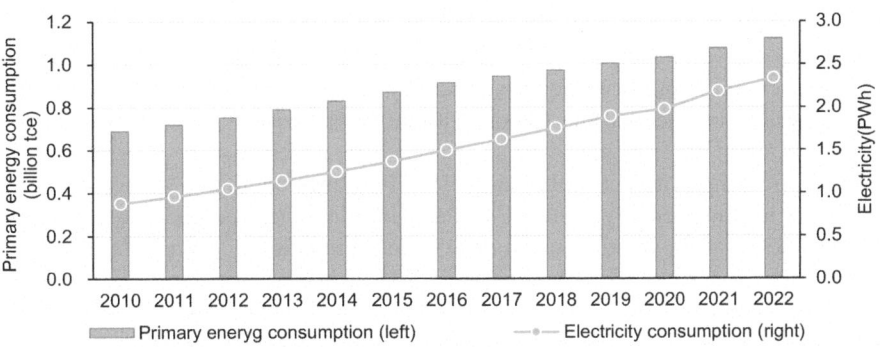

Fig. 1.5 Primary energy consumption and total electricity consumption of building operation in China (2010–2022)

have made their energy consumption the largest proportion of the building energy consumption in China.

Figure 1.7 shows the changes in the total quantity and intensity of the four categories of energy use from 2010 to 2022. The total quantity and intensity in the four categories mainly exhibit the following characteristics:

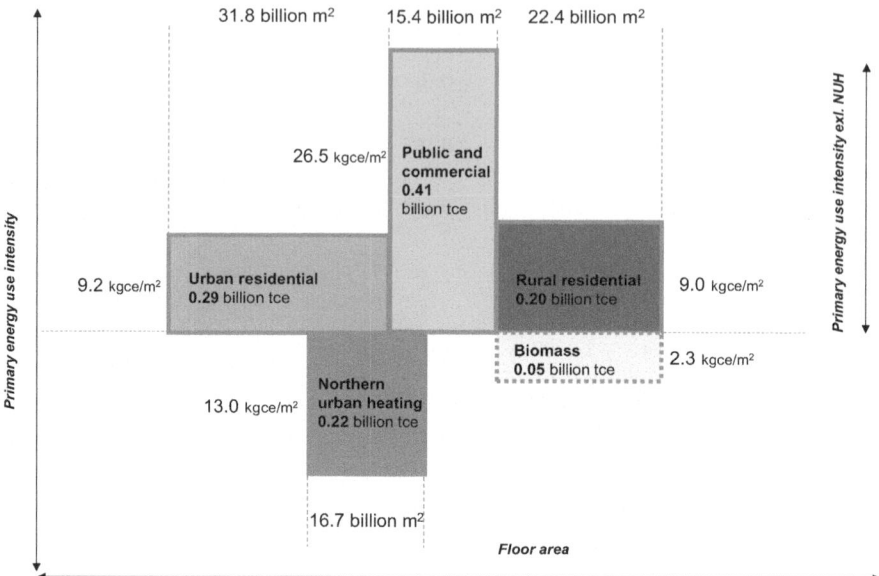

Fig. 1.6 Building operation energy consumption in China (2022). *Note* Electricity, heat, and fuels are uniformly converted into primary energy, which is measured in standard coal, and the electricity consumption is converted into primary energy consumption calculated in standard coal based on the annual average coal consumption for power supply in China. The conversion coefficient in 2022 was 301 gce/kWh

The energy use intensity in NUH was relatively large but has been decreasing with the improvement of the new energy-saving standards and the heat source efficiency in recent years, and the total energy consumption remained stable without any further increase.

Energy use intensity per unit area of P&C buildings continued to increase. The increasing terminal energy demand of P&C buildings (air conditioners, devices, lighting, etc.) was the major cause for the increase in building energy use intensity. In particular, some large new buildings with large-scale centralized systems have been constructed in many cities in recent years, with energy use intensity much higher than that of similar buildings. As the size of P&C buildings grows, their total energy consumption is still increasing.

The energy use intensity per household of urban residential buildings increased because there was an increasing demand for domestic hot water, air conditioners, and household appliances. The issue of heating in winter in the HSCW zone also aroused extensive discussions. There was not too much increase in the energy consumption of lighting in residential buildings because of the adoption of energy-efficient illumination devices. The energy use intensity of cooking also remained unchanged. With the further promotion of urbanization and the growth in the size of urban residential buildings, their total energy consumption is still increasing. The commercial energy use intensity per household of rural residential buildings increased slowly.

1.3 Energy Consumption of China's Building Sector

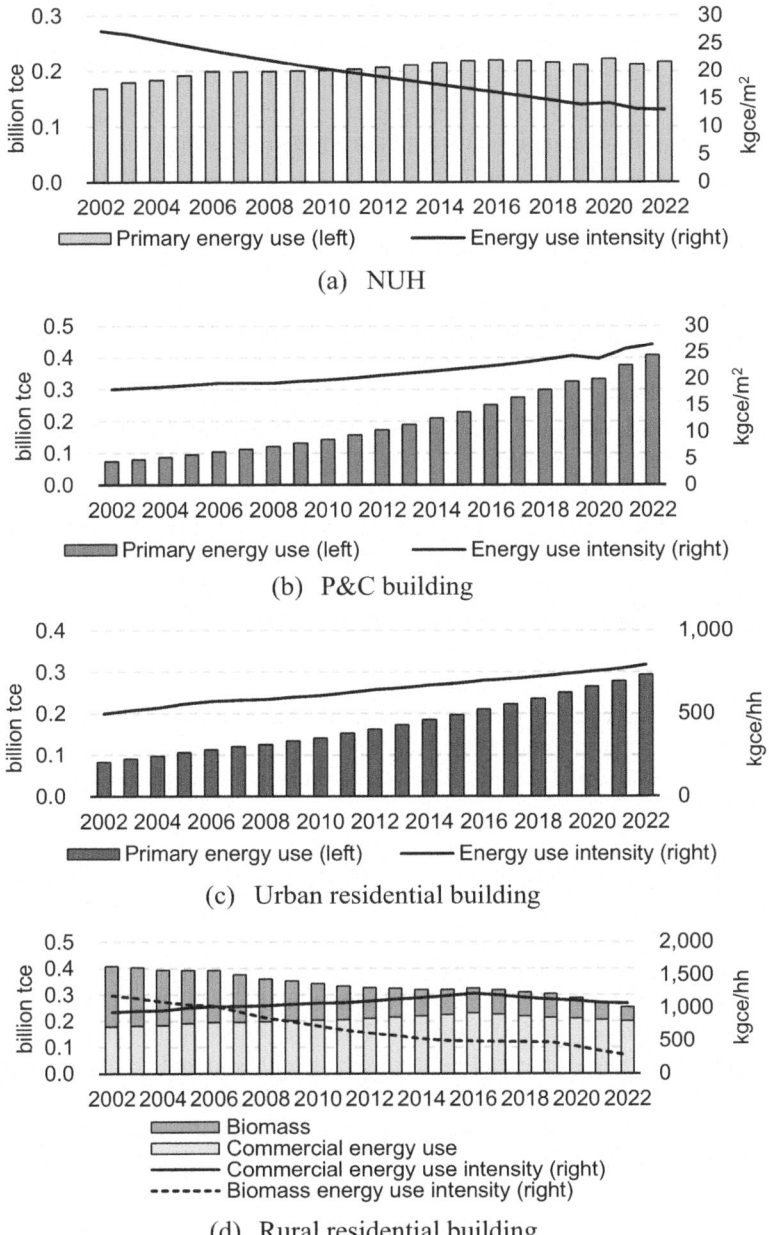

Fig. 1.7 Changes in total building energy use and energy use intensity by year (2002–2022)

As the rural population and the number of households slowly decreased, commercial energy consumption in rural areas remained stable. However, as household appliances became more popular in rural areas and the policy of "switching from coal to electricity" was implemented for clean heating in northern China, electricity consumption has increased dramatically in recent years. Meanwhile, biomass energy use has dropped continuously. Hence, the total energy use of rural residential buildings has declined slowly in recent years.

(1) **NUH**

In 2022, the energy consumption of NUH was 217 million tce, making up 19% of the total energy consumption of buildings in China. From 2002 to 2022, the NUH area tripled from 6.1 to 16.7 billion m^2, while the total energy consumption increased, but did not double. Obviously, the increase in total energy consumption was less than the increase in building floor area, indicating that remarkable results had been achieved in energy saving. The average energy consumption per unit area of heating was 13.0 kgce/m^2 in 2022, a significant decline from 27.3 kgce/m^2 in 2002. Specifically, the main reasons for the decrease in energy use intensity include the improvement of building insulation, which resulted in the decrease in the heat demand of buildings, as well as the increased share of efficient heat sources, and the improvement of operation management. In recent years, the total energy consumption of NUH has been on a declining trend from the peak around 2017. Due to the COVID-19 pandemic, the heating time was extended to varying degrees in different places in the 2019–2020 heating season. According to the statistical data released in the 2023 Annual Report on Urban Heating Development in China, 86% of cities extended the heating time in the 2021–2022 heating season, so the total energy consumption in NUH went up a little in the 2021–2022 heating season.

Gradual improvement of building envelope performance. In recent years, the Ministry of Housing and Urban–Rural Development of the People's Republic of China has adopted various methods to improve building insulation, including the establishment of building energy efficiency design codes that cover different climate zones and building types, the special examination of energy-saving work that started from 2004, and the renovation of existing residential buildings during the 13th Five-Year Plan period. During the 13th Five-Year Plan period, the energy efficiency design standard for new urban residential buildings in severe cold and cold areas in China was raised to "75% energy-saving standard", approximately 10 million square meters of ultra-low and near zero energy buildings completed construction, and 514 million square meters of existing residential buildings and 185 million square meters of P&C buildings completed the energy-saving retrofit. These methods have greatly enhanced building insulation in China and lowered the actual heating demand of buildings, especially in northern China.

Optimization of heat source structure and significant improvement of heat source efficiency. Recent years have seen a gradual increase in the share of efficient CHP to gradually replace boilers. The results of urban heating surveys in 2013, 2016, and 2020 revealed that the proportion of CHP in heat sources for NUH was 42, 48, and 55% respectively in the three years. Gas-fired boilers replaced coal-fired boilers.

1.3 Energy Consumption of China's Building Sector

From 2013 to 2020, the proportion of coal-fired boilers dropped from 42 to 13%, while that of gas-fired boilers increased from 12 to 22%. In the meantime, all types of new heat sources kept on growing, with rising proportions of industrial residual heat, residual heat from nuclear power, ground-source heat pumps, and biomass in heating. Heating system efficiency has also been increasing notably in recent years, thus enabling the overall improvement of the efficiency of all types of centralized heating systems.

(2) **Urban Residential Buildings (Excluding NUH)**

Urban residential building energy consumption (excluding NUH) in 2022 was 291 million tce, accounting for 26% of the total commercial energy consumption in the building sector. Electricity consumption was 630.7 TWh. With the economic and social development and the improvement of living standards in China, the average annual growth rate of urban residential building energy consumption reached up to 7% from 2002 to 2022, and the terminal electricity consumption in 2022 quintupled that in 2002.

From the view of energy use, cooking, household appliances, and lighting consumed the most energy in urban residential buildings (excluding NUH) in China. Thanks to policies and projects for improving the energy efficiency of cooking, household appliances, and lighting, the terminal energy consumption of these three categories was kept under control, and the total energy consumption has undergone a slower increase in recent years. Improving energy efficiency and lowering standby energy consumption should become the optimal methods to limit the energy consumption of cooking, household appliances, and lighting. For example, the promotion of energy-saving lamps significantly improved the lighting efficiency of residential buildings. Energy efficiency standards and behaviors need to be upgraded to lower the electricity consumption from long standby time and frequent reheating and restarting of household appliances. The production standards of such appliances as TV set-top boxes, water dispensers, and electric toilet seats need to be improved to lower the waste of energy when they are in standby mode. The methods, for example, include improving the controllability of set-top boxes, enhancing the insulation capacity of water dispensers, and adopting intelligent control of toilet seats. Policy incentives or subsidies shall not cover electric appliances such as clothes dryers, which may change the lifestyle, and we should watch for the energy spikes of these high energy consumption appliances. Even though the energy consumption of winter heating, summer cooling, and domestic hot water accounts for a smaller proportion in the HSCW zone and the energy consumption per household is at a low level, they have been growing rapidly. The annual average growth rate of heating energy consumption in the HSCW zone could be well over 50%. Therefore, saving terminal energy use for those three categories should be our priority in the next stage of energy saving for urban residential buildings. We should avoid the massive adoption of centralized systems, promote decentralized systems in residential buildings, improve the energy efficiency standards of distributed equipment, and prevent drastic energy consumption increases while improving the indoor service level. Refer to the

China Building Energy Use and Carbon Emission Yearbook 2021 for detailed discussions about the energy-saving and emission-reduction pathways for urban residential buildings in China.

(3) **P&C Buildings (Excluding NUH)**

In 2022, the total floor space of China's P&C buildings was approximately 15.4 billion m^2, and the total energy consumption of P&C buildings (excluding NUH) was 0.408 Gtce, making up 37% of that of the building sector. Electricity consumption stood at 1.27 PWh. The total area of P&C buildings and the proportion of large P&C buildings were all on the rise, which led to an increase in energy demand. The energy consumption per unit area of P&C buildings grew from 18 kgce/m^2 in 2002 to over 26.5 kgce/m^2 in 2022. The energy use intensity increased rapidly, and the total energy consumption surged as well.

In 2020, due to the impact of COVID-19, the operation duration and intensity of public and commercial buildings were subject to pandemic-related control measures, leading to a slight decrease in the average energy use intensity of public and commercial buildings in China. In 2021, the growth rate of energy used in the operation of public and commercial buildings picked up. Since 2001, the newly built building stock of P&C buildings amounted to almost 8 billion m^2, approximately 79% of the current stock. This means that three-fourths of P&C buildings were built after 2001. There are two reasons for this increase. First, a lot of new commercial buildings such as office buildings and commercial complexes have been constructed in recent years. Second, the scale of public service buildings such as schools, hospitals, and sports stadiums has increased due to the necessity for the gradual perfection of relevant infrastructure to promote the building of a well-off society in an all-around way and the improvement of public services. In recent years, the proportion of school and hospital buildings in the new public and commercial buildings has been increasing gradually. In 2021, there were more new hospital and school buildings than new office and hotel buildings.

While the stock of P&C buildings is growing, the number of large-scale P&C buildings is also increasing. In particular, many public and commercial buildings completed in recent years are large top-grade commercial buildings with central air-conditioning. Their electricity consumption per unit area is over 100 kWh/m^2, while the electricity consumption of smaller schools, offices, and stores built in the past is approximately 60 kWh/m^2. The average electricity consumption of public and commercial buildings will continue to increase as the proportion of such new public and commercial buildings with high energy consumption in the total number of public and commercial buildings has been on the rise. Due to the volume and form constraints of such new buildings, the energy use intensity of air conditioning, ventilation, lighting, and elevators in them is much higher than that in general public and commercial buildings. This is also an important cause of the continuous growth in the energy use intensity of public and commercial buildings in China. Refer to the *China Building Energy Use and Carbon Emission Yearbook 2022* for detailed discussions about paths toward energy saving and emission reduction for P&C buildings in China.

1.3 Energy Consumption of China's Building Sector

(4) **Rural Residential Buildings**

In 2022, the commercial energy consumption of rural residential buildings was 0.201 Gtce, accounting for 18% of the total energy consumption of buildings in China. Electricity consumption was 348.1 TWh, and the consumption of rural biomass energy (straw and firewood) was equivalent to about 0.05 Gtce. From 2002 to 2022, urbanization led to a decline of the rural population from 800 to 500 million, and the scale of rural residential buildings was maintained at approximately 23 billion m^2 and has begun to decrease slowly in recent years.

Owing to the higher availability of electricity in rural areas, higher income for rural residents, and more household appliances, the electricity consumption per household in rural areas has increased rapidly. For instance, the number of air conditioners per hundred households in rural areas increased from 16 in 2001 to 89 in 2021, which led to not only the growth of electricity consumption but also a longer peak power load in rural areas during summer. The implementation and promotion of "switching from coal to electricity" in northern China contributed to significant growth in winter heating electricity consumption and peak power load there. Moreover, increasing biomass energy has been replaced by commercial energy, leading to a rapid reduction of the proportion of biomass energy in household energy consumption in rural areas. China released the *Work Plan for Implementation of the PV Poverty Alleviation Project* in 2014, which proposed the development of the photovoltaic (PV) industry in rural areas as an important means of poverty alleviation. A new energy system based on rooftop PV may be built by taking advantage of abundant renewable resources in rural areas. Such a system can realize the net power output to the power grid while meeting the energy demands of rural life, production, and transportation. It can totally cancel the use of fossil and biomass fuels while raising the living standard in rural areas. Thus, it will not only root out the environmental pollution and carbon emission problems caused by the burning of fossil and biomass fuels but also make the production and output of zero-carbon energy another important economic activity in rural areas. Besides, this can make an important contribution to the sustainable development of China's energy system and become an important part of the rural revitalization strategy.

In recent years, with the thorough implementation of haze control measures and clean heating in eastern China, governments at all levels and relevant enterprises have made huge investments to increase the power supply capacity, lay gas pipe networks and change original small household coal-fired furnaces into low-pollution forms in rural areas, which leads to a substantial increase in electricity and gas consumption. The change in rural energy structure will lead to a fundamental transformation of rural energy use patterns, thus facilitating the modernization of rural areas. This opportunity should be leveraged and scientific planning should be made to revolutionize rural energy supply and consumption and establish a new energy system with renewable energy sources as the mainstay for rural residents, which will play an important role in the current energy revolution of China.

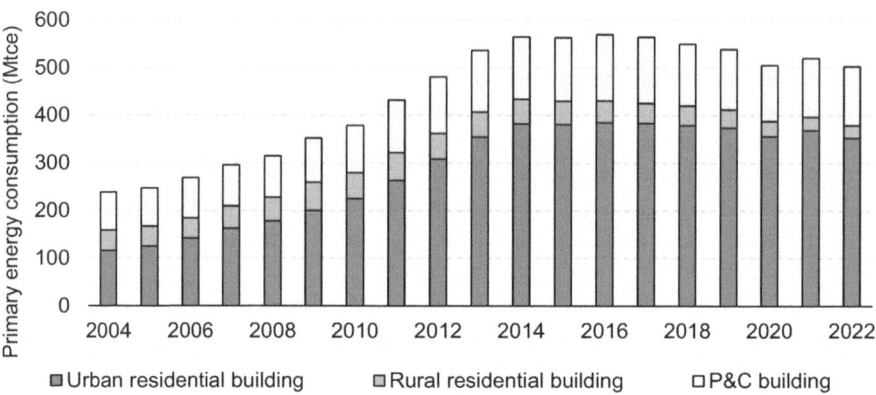

Fig. 1.8 Embodied energy use of China's civil buildings (2004–2022). *Source* Estimation by BERC, Tsinghua University. This figure only covers civil building construction[3]

1.3.2 Embodied Energy Consumption of Building Sector

China's continuous advancement of urbanization in the last two decades has also enabled the embodied energy consumption of civil buildings to become an important part of the total energy consumption of the whole society. A large quantity of building materials is needed, the production process of which lead to great energy consumption and carbon emissions. This is one of the key reasons for the continuing growth of energy consumption and carbon emissions in China.

According to the estimation from the BERC, embodied energy use of civil buildings in China amounted to 0.5 gigatonnes of coal equivalent (Gtce) in 2022, accounting for 9% of China's total energy consumption. The embodied energy use of civil buildings in China has peaked at 2016, while due to the slow decrease in the newly built building stock of civil buildings in recent years, its embodied energy use has also dropped gradually since 2016 as shown in Fig. 1.8. In 2022, the embodied energy of urban residential, rural residential, and P&C buildings accounted for 70, 5, and 25%, respectively.

In fact, the construction sector consists of not only civil buildings but also buildings for production purposes and infrastructures such as motorways, railways, and dams. The embodied energy use of the construction sector mainly includes all types of energy use related to the construction of buildings and infrastructures. According to the calculation of BERC, Tsinghua University, the total embodied energy use of China's construction sector in 2022 was 1.45 Gtce, accounting for up to 27% of the primary energy consumption of the whole society. From 2004 to 2022, embodied energy use in China's construction sector grew from approximately 0.4 to 1.45 Gtce, as shown in Fig. 1.9. Embodied energy from building materials is the mainstay of

---

[3] The newly built building stock data is based on the data under "the statistical standards for construction enterprises as specified in the *China Statistical Yearbook on Construction*".

1.3 Energy Consumption of China's Building Sector

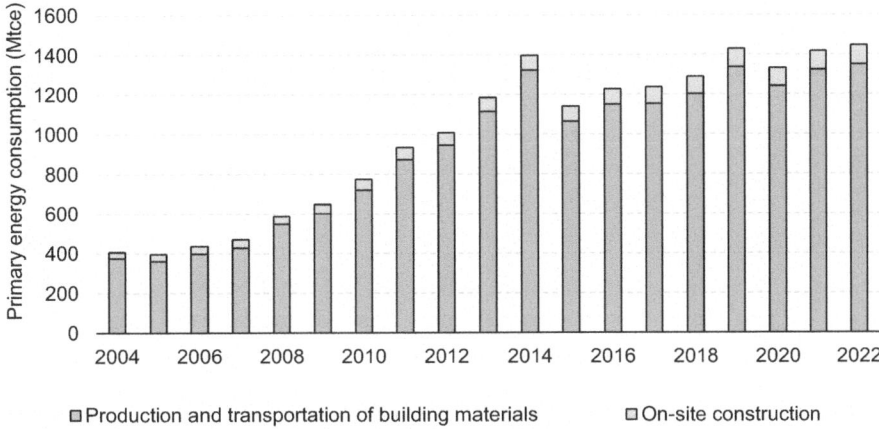

Fig. 1.9 Embodied energy consumption of China's construction sector (2004–2022). *Source* Estimation by BERC, Tsinghua University. *Note* The construction sector involves the construction of civil buildings, production buildings, and infrastructures[4]

the total embodied energy use of buildings, in which iron and steel and cement production consumes more than 80%.

The construction demands of rapid urbanization in China have not only driven the growth of energy consumption directly but also determined China's industrial structure which is dominated by traditional heavy and chemical industries including steel and cement. This is also a key reason for the high energy consumption per unit of industrial value added in China.

There was 1.35 Gtce of industrial energy used for producing building materials in 2022. Between 2013 and 2022, building materials accounted for approximately 40% of the total industrial energy consumption, as shown in Fig. 1.10. The fast pace of urbanization in China drove the demand for building materials, which was the key reason why the higher percentage of total energy consumption was represented by iron and steel, building materials, and other traditional heavy industries.

As urbanization and infrastructure development have achieved initial progress in China, the transformation of the construction mode is ongoing. In 2022, the per capita floor area for urban residential buildings was 35 m^2 in China, which was close to that of some developed countries in Asia, such as Japan and South Korea, but still far lower than that of the US. The reason was that during China's urbanization, the main type of building in the urban communities was apartments instead of single houses, such as that of the US. From the perspective of urban form, the utilization ratio of P&C buildings was high in China because of the high-density large city mode of development, so it is unnecessary to follow the per capita P&C building scale in Europe and America. In the future, there will be no more rapid growth of iron and steel, building materials, and other high energy consumption industries, so long

---

[4] Data about the consumption of building materials is from the *China Statistical Yearbook on Construction*.

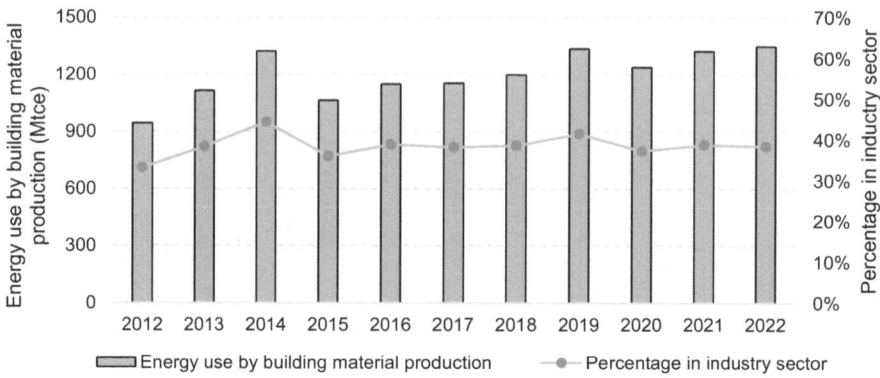

Fig. 1.10 Energy consumption for building material production[5]

as there is less demolition, and the building life cycle can be properly maintained. Therefore, for the next round of urbanization, the demolition of buildings that have not reached the end of life shall be abolished. Technologies to extend the building life cycle shall be invented. Buildings and infrastructures should be properly repaired, and the building life cycle should be extended to facilitate industrial transformation and total energy control.

1.4 GHG Emissions of China's Building Sector

1.4.1 Carbon Dioxide Emissions from Building Operation

The carbon dioxide emissions from building operation are affected by the growth in total energy demand of buildings, the improvement of building energy efficiency, and the adjustment of building fuel types and energy supply structure. Electricity, coal, and gas were the major energy sources for building operation. Electricity accounted for 65% of the total energy use in urban residential buildings and P&C buildings, in which CO_2 was indirectly emitted. The adoption of CHP in NUH could also lead to indirect CO_2 emissions. The percentage of coal and gas consumption was higher than that of electricity consumption for NUH and rural residential buildings. The percentage of coal and gas consumption was about 89% for NUH and the percentage of fossil energy consumption was about 48% for rural residential buildings, which led to massive direct carbon dioxide emissions. In another aspect, as the percentage of zero-carbon electricity has increased in China, the average emission factors have declined tremendously, at 541 gCO_2/kWh in 2022. Besides, the share of electricity consumption in building operation energy consumption gradually increased as well.

---

[5] Materials for the construction sector here mainly include steel, cement, aluminum, glass and architectural ceramics.

1.4 GHG Emissions of China's Building Sector

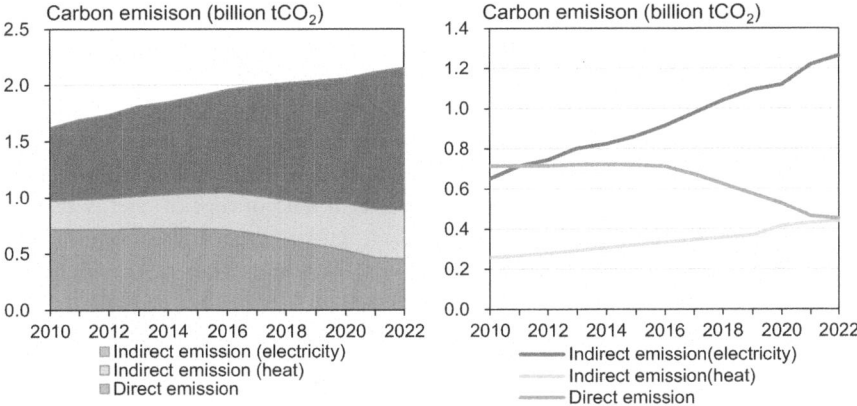

Fig. 1.11 Carbon dioxide emissions from building operation (2022)

These two trends have promoted the low-carbon development of building operation energy consumption.

According to the analysis results from CBEEM, in 2022, the total carbon emissions during building operation in China were 2.2 billion tonnes of carbon dioxide (Gt CO_2), equivalent to 1.52 t per capita carbon emissions and 31 kg/m² of average carbon emissions per unit area. In the total carbon emissions, the proportions of direct carbon emissions (0.45 Gt CO_2), indirect carbon emissions from electricity use (1.26 Gt CO_2), and indirect carbon emissions from heating (0.44 Gt CO_2) were 21, 58, and 21%, respectively, as shown in Fig. 1.11.

(1) **Direct Carbon Emission**

In 2022, the direct carbon emissions of buildings were 450 million tons of CO_2, including about 150 million tons of CO_2 directly emitted from urban and rural cooking, about 210 million tons of CO_2 emitted from household gas- and coal-fired heating boilers,[6] and 90 million tons of CO_2 directly emitted from natural gas consumption for hot water, steam boilers, absorption refrigeration, and other purposes. Emissions from rural areas accounted for more than half of the direct carbon emissions.

In recent years, with the vigorous promotion of "switching from coal to electricity", "switching from coal to gas" and clean heating in rural areas, the direct carbon emissions in the building sector of China have been declining slowly from the peak around 2015. Direct carbon emissions of the building sector will continue to decrease and no longer reach a new peak as long as electrification is continuously promoted in new buildings.

---

[6] Refers to gas- and coal-fired heating boilers installed in urban and rural residential buildings as well as coal- and gas-fired boilers installed in public and commercial buildings. Such fuels are burned directly in buildings, so the resulting carbon emissions fall into the category of direct carbon emissions from buildings.

To reduce direct carbon emissions of the building sector to zero, the key lies in the point-in-time and intensity of promoting "electrification". It is expected that direct carbon emissions of buildings will be reduced to zero during 2040–2045. Analyses indicate that electrification will not increase the operation cost in 80% of all cases and that the initial investment in equipment will be returned in about 5 years through the reduction of operating cost. Therefore, the major obstacle to promoting electrification in buildings is not economic cost but the change in the concept of energy use and cooking culture. Increasing the publicity of "zero carbon emissions of buildings by electrification" among the public and promoting "switching from gas to electricity" in new and existing buildings are the most important approaches to realizing zero direct carbon emissions from building operation.

(2) **Indirect Carbon Emission from Electricity Use**

In 2022, the electricity consumption by building operation in China was 2.3 PWh, and the indirect carbon emission from electricity use was 1.26 billion tons of CO_2. At present, the per capita electricity consumption of the building sector in China is one-sixth of that in the US and Canada and about one-third of that in France and Japan, while the electricity consumption per unit area of buildings in China is one-third of that in the US and Canada. The difference in lifestyle and building operation is one of the main reasons for the difference in electricity use intensity between China and developed countries.

In recent years, the carbon emission increase caused by the growth in electricity consumption of buildings has exceeded the carbon emission decrease resulting from the reduction of the carbon emission factor of electricity. The indirect carbon emissions from electricity consumption of buildings will keep increasing before peaking. China should maintain a green and economical lifestyle and building operation mode to avoid any surge in energy use of buildings that once occurred after rapid economic growth in the history of the US, Japan, and other developed countries. When the building floor area in China reaches 75 billion m^2 in 2060, the electricity consumption of buildings should be 3.8 PWh, which will meet the demand of the Chinese people for a good life and energy consumption of buildings. On this basis, the new type of electric power system adopting "photovoltaic, energy storage, direct current and flexibility (PEDF)" technologies should be promoted. When the reduction in indirect carbon emissions from building electricity use caused by the increase of "green electricity" in practical electricity consumption through flexible electricity use is greater than the growth in indirect carbon emissions from building electricity use caused by the growth in the total scale and electricity use intensity of buildings each year, the indirect carbon emissions from the electricity use of buildings in China can peak. Upon overall popularization of the "PEDF" power distribution mode and flexible means of electricity use, the zero-carbon goal for electricity consumption of buildings can be achieved earlier than that for the national power system.

(3) **Indirect Carbon Emission from Heating**

In 2022, the building floor area of NUH in China was 16.7 billion m^2, and the indirect carbon emission from heating during building operation was 440 million tons of CO_2. Recent years have seen continuous growth in the centralized heating area and heating demand in northern China but a continuous decrease in energy consumption and carbon emissions per square meter of heat supply. The total indirect carbon emissions from NUH showed a trend of slow growth. The indirect carbon emissions from heating during building operation can peak by around 2025 through further strengthening the renovation of existing buildings, fully exploiting low-grade residual heat resources, and phasing out scattered coal-fired boilers. Afterward, with the gradual completion of the zero-emission transformation of the remaining thermal power (to be replaced with CCUS and biomass fuel) by the electric power sector, the indirect carbon emissions from building heating can be reduced to zero in sync with the decarbonization process of the power system.

To this end, it is necessary to continue improving the envelope performance of new buildings and the renovation of existing buildings in a strict manner, so that the average heating demand of buildings in northern China can decrease from 0.37 GJ/m^2 at present to below 0.25 GJ/m^2. During 2020–2035, the return water temperature should be lowered through the terminal renovation of centralized heating systems, thus effectively recovering the residual heat of thermal power plants and the industrial low-grade residual heat. The increasing building heating demand will be met by tapping into the potential heating capacity of existing heat sources. CHP transformation will be carried out for the coastal nuclear power in northern China to supply heat to areas within 200 km from the normal of coastal northern China. From 2035, in coordination with the schedule of the thermal power shutdown of the electric power system, seasonal heat storage projects will be implemented simultaneously to solve the problem of reduced heat source power resulting from the shutdown of thermal power plants. Until 2045, the annual residual heat of nuclear power, the annual residual heat of peak shaving thermal power, the residual heat of wind and solar curtailments at the centralized wind and solar PV power bases, as well as the heat emitted annually from the industrial low-grade residual heat will have been collected through seasonal heat storage projects. In this way, the indirect carbon emissions from building heating can be reduced to zero, while zero-carbon emission of the power system is realized.

In view of the four categories, the scale, intensity, and total quantity of carbon emissions from them are represented in the block diagram in Fig. 1.12 respectively, in which the horizontal axis stands for the building floor area and the vertical axis for carbon intensity per square meter for the four categories. The total area of the four blocks represents the total carbon emissions. The growth in the total carbon emissions from the four categories is shown in Fig. 1.13. It can be seen that the characteristics of carbon emissions from the four categories are not the same as those of their energy consumption. As public and commercial buildings have the highest energy use intensity, their carbon intensity per unit area is also at the peak. In 2022, their carbon intensity was 47.5 kg CO_2/m^2, and their total carbon emissions were still on the rise as their total energy consumption and energy use intensity were growing steadily. For NUH, its carbon intensity was only second to that of public and commercial buildings due to the consumption of a lot of coal. In 2022, the carbon intensity was 29.2 kg CO_2/m^2, and the carbon emissions peaked and remained stable

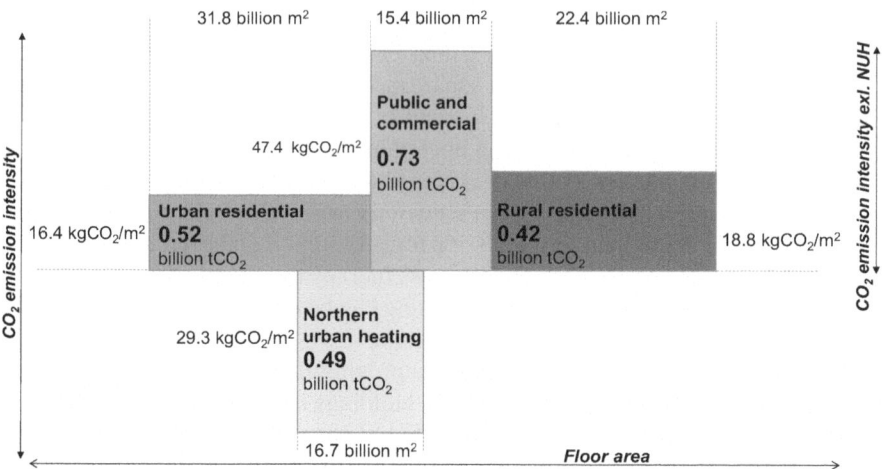

Fig. 1.12 Carbon dioxide emissions from building operation in China (2022)

at approximately 0.5 Gt CO_2 due to the increasing heat demand, the improvement of heat supply efficiency, and the consistent speed in the transformation of energy structure. Although there is little difference in primary energy use intensity per square meter between rural and urban residential buildings, the carbon intensity per square meter of rural residential buildings is higher than that of urban residential buildings due to a low level of electrification and a high proportion of coal fuel. The carbon intensity per unit area of rural residential buildings was 18.5 kg CO_2/m^2. Due to the implementation of "switching from coal to electricity" and "switching from coal to gas" in rural areas, the total carbon emissions of rural residential buildings have peaked and then declined annually in recent years. The carbon intensity per unit area of urban residential buildings was 16.2 kg CO_2/m^2 and was increasing slowly with the growth in electricity consumption.

1.4.2 Embodied CO_2 Emission of Building Sector

As urbanization continues, the embodied energy use of civil buildings in China also increases rapidly. The construction of buildings and infrastructures not only consumes a colossal amount of energy but also leads to a lot of carbon dioxide emissions. In addition to carbon dioxide emissions resulting from energy consumption, emissions in the cement production process[7] are also an important part.

In 2022, the total carbon emissions from civil building construction in China were about 1.5 Gt CO_2, mainly including carbon emissions from energy use for

---

[7] Refers to carbon emissions from chemical reactions (excluding combustion) for cement production.

1.4 GHG Emissions of China's Building Sector

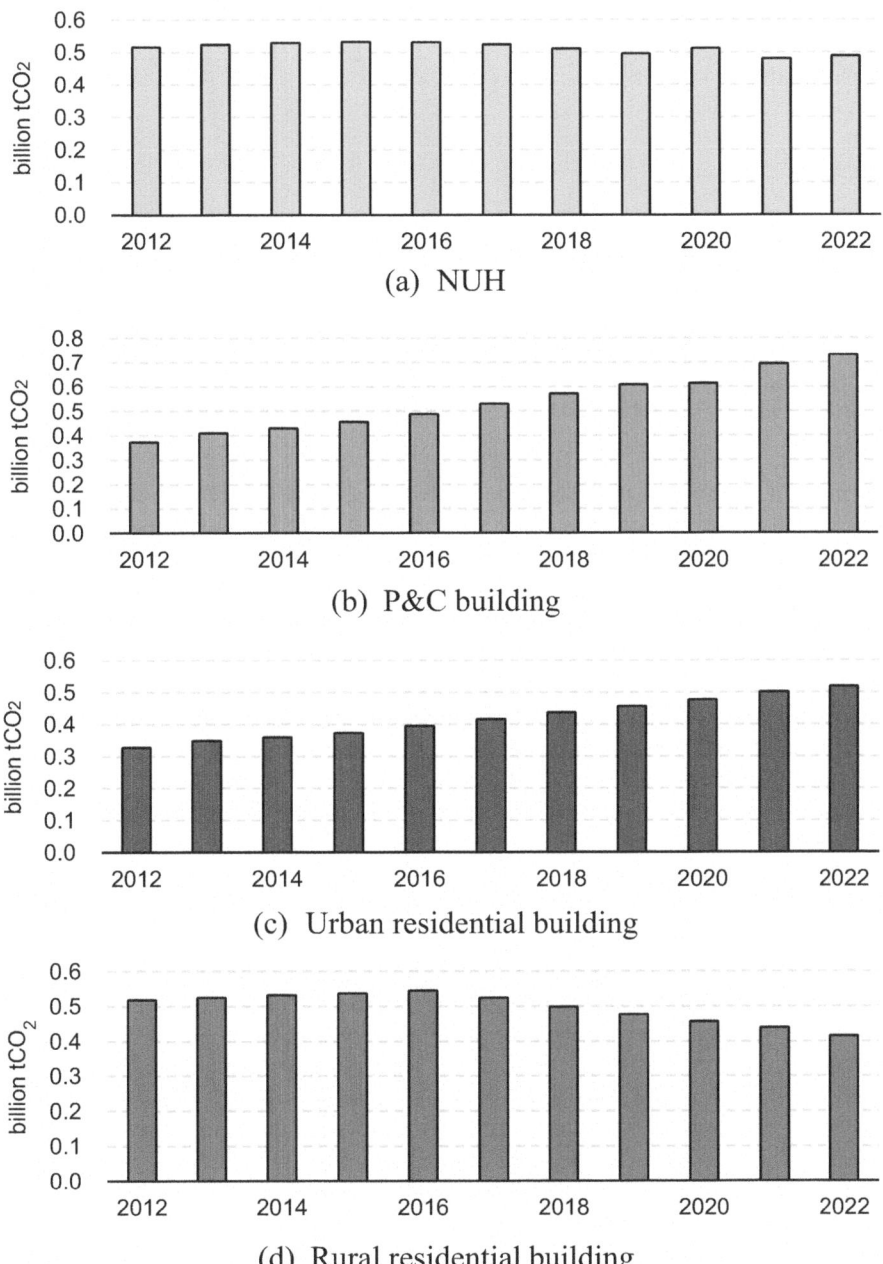

Fig. 1.13 Carbon emissions of building energy use (2012–2022)

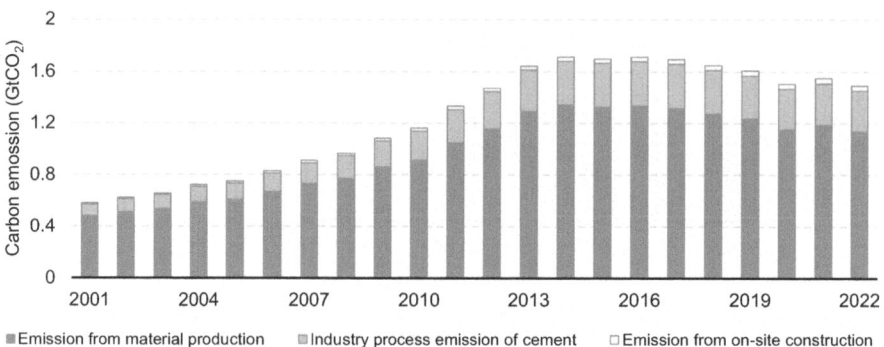

Fig. 1.14 Embodied carbon emissions of civil buildings in China (2004–2022). *Source* Estimation by BERC, Tsinghua University. *Note* Civil building construction is included only

the manufacturing and transportation of building materials (76%), the industrial process emissions of cement (21%) and energy use during construction (3%), as shown in Fig. 1.14. Although such carbon emissions are included in the industry and transportation sectors, they are driven by the demand of the building sector. Hence, the building sector shall also be responsible for such carbon emissions and reduce its demand to contribute to emission reduction. With the end of the large-scale construction period in China, the scale of newly built building stock has been decreasing each year. The carbon emissions from civil building construction peaked in 2016 and have been declining slowly year by year in recent years.

In fact, as China is still in the urbanization stage, various infrastructures need to be constructed in addition to civil buildings. In 2022, the total carbon emissions from construction in the construction sector were about 4.2 $GtCO_2$, nearly half of China's total carbon emissions, as shown in Fig. 1.15. Carbon emissions from civil building construction were about 35% of the total carbon emissions from construction in China's construction sector.

To realize zero carbon emissions from building construction as early as possible, the total number and scale of buildings should be controlled reasonably first to minimize excessive construction and avoid large-scale demolition and construction. At present, the total building stock and per capita floor area in China have met the demands of urban and rural residence, production, and living. By 2060, the production and living demands of the future urban and rural populations in China can be met with 40 m^2 of per capita residential floor area, 15.5 m^2 of per capita P&C building stock, and 75 billion m^2 of total floor area. To realize zero-carbon emissions from building construction in China, the construction speed, total quantity, and scale of buildings need to be planned rationally.

In the meantime, China's construction sector will shift from the large-scale construction of new buildings to the maintenance and functional improvement of existing ones. The construction of houses in China has shifted from increasing the supply of houses to meet immediate needs to demolishing old ones and building new

1.4 GHG Emissions of China's Building Sector

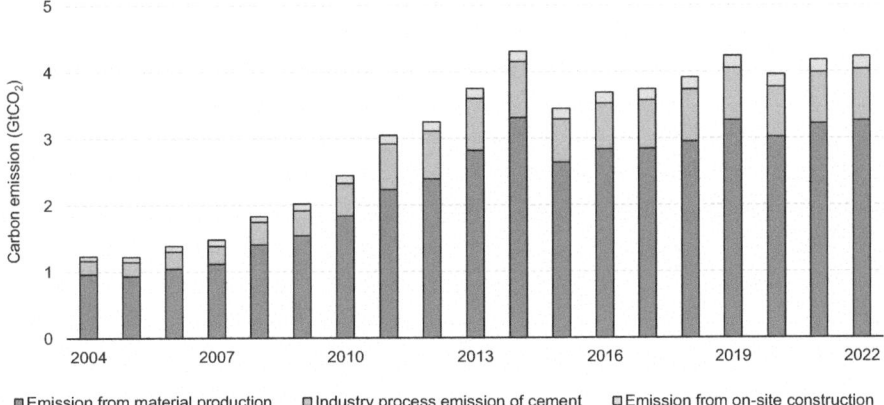

Fig. 1.15 Embodied carbon emissions of the construction sector in China (2004–2022). *Source* Estimation by BERC, Tsinghua University. *Note* The construction sector involves the construction of civil buildings, production buildings, and infrastructures

ones to improve building performance and functionality. The "Mass-Demolition-for-Mass-Construction" pattern has become the main mode of the construction sector. Based on the objective of total number planning for civil buildings in the future and in view of the reasonable construction speed, the gradual transformation from "Mass-Demolition-for-Mass-Construction" to "replacing demolition with fine repair" will enable the stabilization of China's construction sector and the gradual reduction of carbon emissions from civil building construction to 0.2 Gt CO_2. Zero emissions from building construction are expected to be realized in 2050 through further application of new building materials and new structural systems and technologies.

Open Access This chapter is licensed under the terms of the Creative Commons Attribution-NonCommercial-NoDerivatives 4.0 International License (http://creativecommons.org/licenses/by-nc-nd/4.0/), which permits any noncommercial use, sharing, distribution and reproduction in any medium or format, as long as you give appropriate credit to the original author(s) and the source, provide a link to the Creative Commons license and indicate if you modified the licensed material. You do not have permission under this license to share adapted material derived from this chapter or parts of it.

The images or other third party material in this chapter are included in the chapter's Creative Commons license, unless indicated otherwise in a credit line to the material. If material is not included in the chapter's Creative Commons license and your intended use is not permitted by statutory regulation or exceeds the permitted use, you will need to obtain permission directly from the copyright holder.

Chapter 2
Comparison of Energy Consumption and Carbon Emissions from Building Operation Between China and Other Countries

2.1 Energy Consumption and GHG Emissions in the Global Building Sector

2.1.1 Energy Consumption by Global Building Operation

According to the International Energy Agency's (IEA) calculation of global energy use and emissions of the building sector (as shown in the figure below), in 2021, the embodied energy consumption during the global building construction stage (including building and infrastructure construction) and the building operation energy consumption accounted for 37% of the total global energy consumption, in which the embodied energy consumption during building and infrastructure construction accounted for 7% and the energy consumed during the operation stage accounted for 30%. In 2021, the total global CO_2 emissions (including energy-related and industrial process emissions) were 36.3 Gt CO_2, in which the embodied CO_2 emissions from construction (including building and infrastructure construction) in the construction sector accounted for 12% and the CO_2 emissions from building operation accounted for 28% (Fig. 2.1).

According to BERC's calculation of China's building energy use and emissions, in 2022, China's embodied energy use and operational energy use in the building sector accounted for 30% of the total social energy use,[1] which was close to the global level. However, China's building embodied energy use was 9% of the total social energy use, higher than 7% of the global level. If coupled with the embodied energy use for production building and infrastructure construction, China's building embodied energy use will be up to 27% of the total social energy use. The building

---

[1] The primary energy consumption method is adopted for conversion. The heat consumption and electricity consumption of buildings are converted by the coefficient of coal consumption for thermal power supply into primary energy consumption, which is then added together with the consumption of other energy types at terminals.

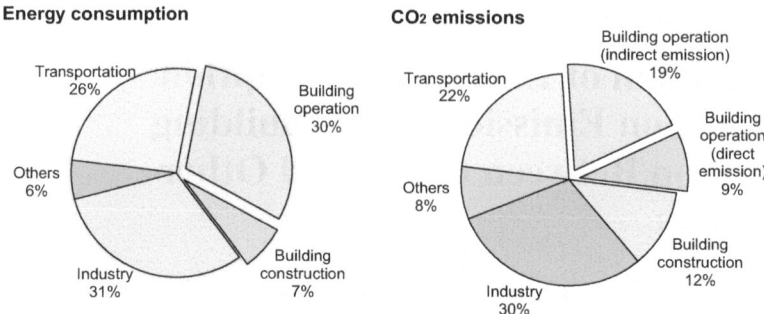

Fig. 2.1 Energy use and CO_2 emissions in the global building sector (2021). *Source* 2022 Global Status Report for Buildings and Construction, International Energy Agency. *Note* The construction sector involves civil building construction, production building construction, and infrastructure construction. This figure uses the terminal energy consumption data provided by IEA, which is obtained through the direct addition of heat consumption for heating, electricity consumption of buildings, and terminal use of various energy types. The electricity consumption is converted to primary energy using a calorific value equivalent method. This conversion method is different from that used in the comparison of building energy consumption among countries in the ensuing part of this research, so it should be treated differently in data comparison

operation energy use accounted for 21% of the total and was still lower than the global average. In the future, the share of China's building sector energy use will continue to increase with economic and social development and the improvement of living standards.

In terms of CO_2 emissions, in 2022, China's total social carbon emissions (including energy-related and industrial process emissions) were about 11.4 Gt CO_2, in which the embodied CO_2 emissions from building construction and the CO_2 emissions from building operation accounted for approximately 32%, including 13% from building construction and 19% from building operation (Fig. 2.2). If the energy-related CO_2 emissions were considered only, the CO_2 emissions from energy consumption of the whole society in China could be about 10.5 Gt CO_2 in 2021, including about 21% from building operation.

As China's urbanization is still ongoing, the major share of energy consumption and emission has come from building and infrastructure construction. The share of embodied energy consumption from building construction in China is higher than the global average and also higher than that of member countries of the Organization for Economic Co-operation and Development (OECD) that have already been urbanized. However, compared with those in OECD member countries, the building operation energy consumption and carbon emissions in China are at a lower level. As urbanization becomes slower in China, the proportions of building operation energy consumption and related emissions in the social total will further increase. China will gradually shift the focus of its building energy efficiency and low-carbon work from the low-carbon development of new buildings to the low-carbon operation of existing ones.

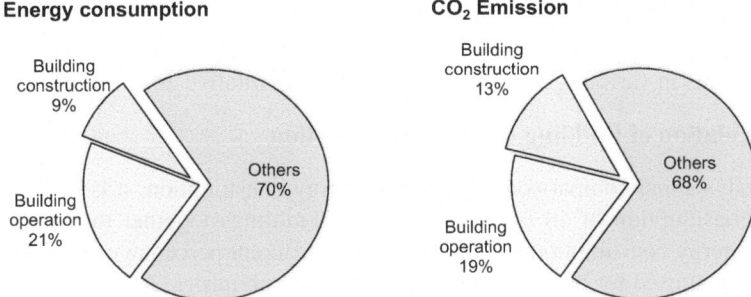

Fig. 2.2 China's building energy use and CO_2 emissions (2022). *Source* Estimation with CBEEM of BERC, Tsinghua University. *Note* The construction sector involves the construction of civil buildings, production buildings, and infrastructures. The diagram on the right illustrates the structure of China's total social carbon emissions (including energy-related and industrial process emissions)

2.1.2 Boundaries and Comparative Study Methods of Building Energy Consumption and Emissions

Comparing building energy consumption among countries is an important means to know the building energy consumption in China, analyze its future development trend, and design the paths of building energy saving. In this section, the building operation energy consumption and carbon emission data of countries in the world are collected and analyzed through comparison. To ensure data comparability and better reflect actual energy use, the energy consumption data in this study only included commodity energy and did not include biomass energy that was not circulating.

Two types of data are collected to compare building energy consumption among countries. The first type includes population, the number of households, and the floor area of buildings. The other is building energy consumption data, which mainly includes the total consumption of electricity, heat, coal, natural gas, and other fuels during building operation. The building energy data of countries around the globe collected in this study mainly comes from two sources:

(1) Databases of international organizations and agencies: mainly including IEA, Odysee, World Bank, and Eurostat databases;
(2) Official statistics of countries: For example, Japan's data mainly comes from the *Statistical Handbook of Japan* and the *Japan Statistical Yearbook* published by the Statistics Bureau of Japan. The data on the US is mainly sourced from the periodic surveys conducted on representative buildings of the country and the statistics released per year by the Energy Information Administration (EIA). Canada's data mainly comes from Natural Resources Canada. The data on South Korea mainly comes from the building information statistics of the Ministry of Land, Infrastructure, and Transport and the KOSIS data. The data of India is mainly from the National Statistical Office (NSO) and the Ministry of Statistics and Programme Implementation (MoSPI).

(3) Some published research reports and literature also provide important support and reference for this study as they have studied the building energy and emissions in various countries and provided quantitative data.

(1) **Calculation of Building Energy Consumption**

In the analysis and comparison of building energy consumption, it is necessary to add the consumption of all types of energy in buildings together to get the total building energy consumption, on account of the different percentages of electricity, fuels, and heat used for building operation in various countries. At present, the end-use energy consumption method and the primary energy consumption method are commonly used for calculation. In the context of the low-carbon energy transition, the development trend of building energy use in various countries is to achieve full electrification. With the gradual increase in the percentage of electricity in the energy structure, it will be more meaningful to convert all categories of energy use into electricity and then add them together to get the total energy consumption of buildings. Therefore, the total energy consumption of buildings is calculated in this section through the conversion of various types of energy into electricity. To decouple the level of building energy use and the level of the energy conversion system, a unified energy conversion factor benchmark is used for the conversion between each fuel and electricity in this study. Under the principle of the energy conversion factor benchmark, the total global building energy use can be directly allocated to the global primary energy, and the energy conversion systems have positive and negative values respectively to reflect their efficiency (high or low) and energy structure (good or bad), and the sum is zero. The energy conversion factor benchmark theoretically means the global average conversion level, namely the global average of the power generation capacity of each fuel. The energy conversion factor benchmark used in this section are listed in Table 2.1.

(2) **Calculation of Carbon Emissions from Buildings**

The data on carbon emissions from building operation of different countries in this section is sourced from IEA and the calculation results of the CBEEM model by BERC, Tsinghua University. When calculating the total carbon emissions from building operation, the direct carbon emissions, indirect carbon emissions from electricity use, and indirect carbon emissions from heating in buildings were considered. When calculating indirect carbon emissions from building electricity use, the total

Table 2.1 Energy conversion factor benchmark in the calculation of total energy consumption of buildings in various countries

Energy	Unit	Conversion factor benchmark
Coal	gce/kWh	300
Oil	goe/kWh	191
Natural gas	Nm^3/kWh	0.2
Heat from boiler	kWh/GJ	133
Heat from CHP	kWh/GJ	70

carbon emissions from electricity generation in each country were divided by the total electricity generation to obtain the average carbon emission factor for electricity use in each country. The carbon emission factor was used to calculate indirect carbon emissions from building electricity use. Carbon emissions from building heat use were calculated with the building heat use and the carbon emission factor per unit of heat. In the study of the building operation energy consumption in each country, various types of energy were converted into electricity by taking a unified conversion factor benchmark as the conversion coefficient. In the study of carbon emissions from building operation in each country, the real carbon emission factor instead of a unified carbon emission factor was used to calculate the real carbon emissions because carbon emissions from buildings are closely related to the energy structure and must be discussed together with the energy structure and the energy conversion system.

For carbon emissions during building operation, each country proposed a goal to reduce carbon emissions in the building sector. The technical pathways and priorities to achieve carbon neutrality in buildings differ between countries. To quantify and analyze the various problems confronted by each country in their attempt to achieve carbon neutrality in the building sector, the carbon emission factors for electricity and heat of each country were used in the calculation. Therefore, differences in the energy mix and efficiency among countries will affect the total amount and intensity of carbon emissions during building operation.

2.2 Energy Consumption and Carbon Emissions in the Building Sector of Different Countries

2.2.1 Building Operation Energy Consumption of Different Countries

Three indicators were selected to compare building energy use across countries: total amount, energy use per capita, and energy use per floor area, as illustrated in Fig. 2.3. The building energy use in this figure was calculated by adopting the electricity-equivalent method to calculate the total energy use of building operation. Energy use intensity per capita and energy use intensity per floor area are shown on the horizontal and vertical axes, respectively. The size of the bubbles represents the total energy use for building operation in specific countries. The bubble chart demonstrates that total energy use for building operation in China was similar to that in the US, although the energy use intensity remained at a lower level. The energy consumption per capita and per floor area in China were far lower than those in the US, Canada, Europe, Japan, and South Korea. The equivalent electricity consumption per capita for building operation was about one-fifth of that in the US and Canada and about half that in Japan and South Korea. The equivalent electricity consumption per floor area for building operation was one-third of that in Canada and half that in the

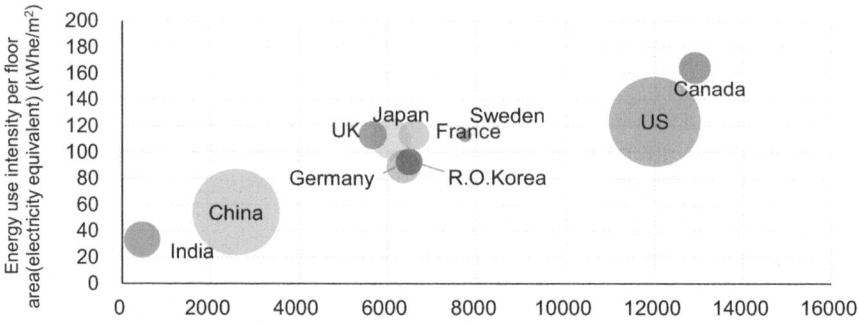

Fig. 2.3 Comparison of building operation energy consumption in different countries (electricity equivalent method). *Source* CBEEM of BERC, Tsinghua University; World Energy Balances, Energy Efficiency Indicators database (2023 edition) of IEA; WDI database of the World Bank; Satish Kumar (2019) of India.[2] Data from 2020 for Canada, and data from 2020 for other countries

US, Europe, Japan, and South Korea. In the context of tackling climate change and reducing carbon emissions, most countries are carrying out energy transformations, including promoting electrification in the building sector and replacing fossil fuel-based energy with renewable electricity. China needs to develop a different route from developed countries to achieve the target for low carbon emission and energy-saving in the building sector. This would pose a significant challenge to China's low-carbon and sustainable development in the building sector. Meanwhile, many developing countries are experiencing rapid changes in building energy use. China's building energy development pathway will serve as an important reference for many countries' choices, which will further influence global building energy development.

Comparing the proportion of electricity in the final energy consumption of the construction sector across various countries from 2001 to 2021, as shown in Fig. 2.4. Sweden, the United States, Japan, and Canada have consistently maintained a relatively high level of electrification in the construction sector, exceeding 40% since the beginning of this century and still maintaining a steady growth trend. France and South Korea have experienced rapid growth in electrification rates, rapidly increasing from around 30% in 2001 to now surpassing 40%. The electrification rates in the construction sectors of the United Kingdom and Germany have remained relatively stable, with slow growth mainly due to these two countries still retaining a certain proportion of fossil energy for building heating. China's electrification rate in the construction sector has rapidly increased from 17% in 2001 to 38% in 2021, surpassing both the United Kingdom and Germany, and is currently in a phase of rapid growth.

---

[2] Satish Kumar et al. (2019). Estimating India's commercial building stock to address the energy data challenge. Building Research & Information, 2019, 47, 24–37.

2.2 Energy Consumption and Carbon Emissions in the Building Sector ...

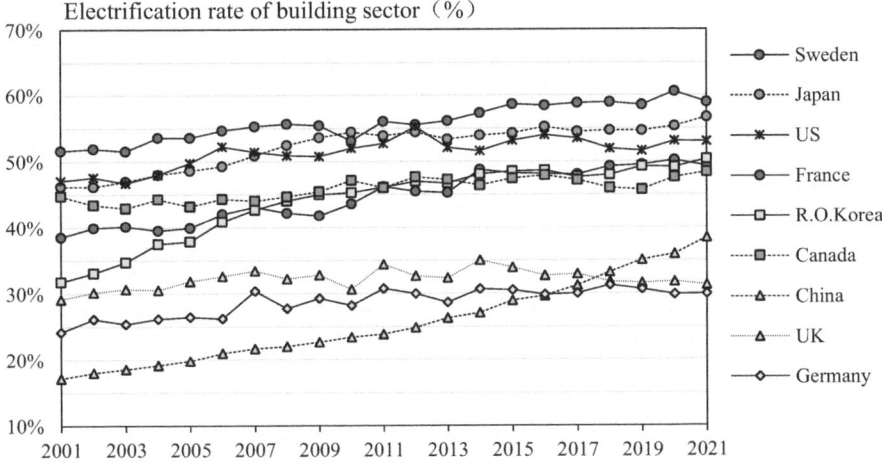

Fig. 2.4 The electrification rate of energy use in the construction sector by country (2001–2021)

2.2.2 Carbon Emissions from Building Operation in Different Countries

The per capita total carbon emissions and the proportion of carbon emissions from the construction sector by country are shown in Fig. 2.5. From the figure, it can be seen that China's current per capita total carbon emissions (including industrial, construction, transportation, and electricity sectors) are slightly higher than the global average but still lower than countries like the United States and Canada. In terms of per capita carbon emissions from building operations, it is also slightly higher than the global average but significantly lower than developed countries. This is mainly because China is still in the process of industrialization and urbanization,

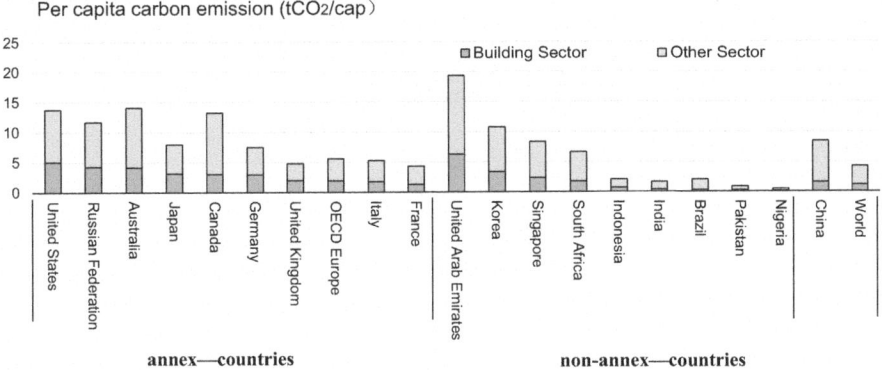

Fig. 2.5 Per capita carbon emissions comparison by country (2021)

and the proportion of carbon emissions from building operations to the total societal carbon emissions is still lower than that of developed countries. In recent years, China's pressure to address climate change has been increasing, and the construction sector also needs to achieve low-carbon development and peak emissions as soon as possible. How to achieve this goal is another major challenge for the development of the construction sector.

Several countries have set their own goals for achieving carbon neutrality and their paths to achieving carbon neutrality in the building sector. Reducing carbon emissions in the building sector is also one of the important fields to realize carbon neutrality in the whole society. Figure 2.6 shows the total carbon emissions from building operation (bubble chart area), per capita carbon emissions (horizontal axis), and carbon emissions per unit building floor area (vertical axis) of different countries, which are converted according to the energy structures of these countries. The bubble chart of carbon emissions demonstrates that carbon emissions in the building sector are affected not only by the total energy consumption but also by the energy structures of these countries. The per capita carbon emissions and carbon emissions per unit floor area from building operation in China are lower than those in most developed countries due to China's low building operation energy consumption. However, the energy structure of France is dominated by low-carbon nuclear power; although it has higher building energy use intensity than China does, its carbon intensity is lower than that of China. This also shows that the low-carbon transition of both energy systems and building energy use structure should be achieved in addition to the improvement of energy saving and energy efficiency of buildings on the path toward carbon neutrality.

In addition to the comparative analysis of carbon emissions from building operations in various countries, analyzing the trends in carbon emissions is also crucial. Figure 2.7 compares the trends in carbon emissions from building operations from

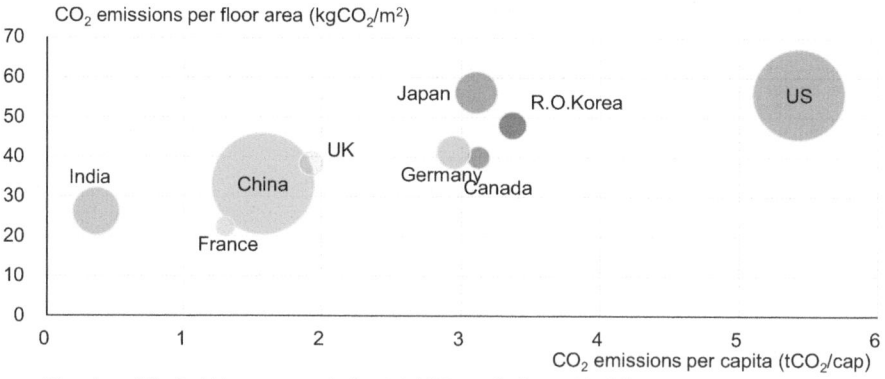

Fig. 2.6 Comparison of per capita carbon emissions of different countries (2021). *Source* Data of countries in 2021 as provided in the CO_2 Emissions from the Fuel Combustion Highlights 2023 database, IEA. Data from China are the results of CBEEM of BERC, Tsinghua University

2.2 Energy Consumption and Carbon Emissions in the Building Sector …

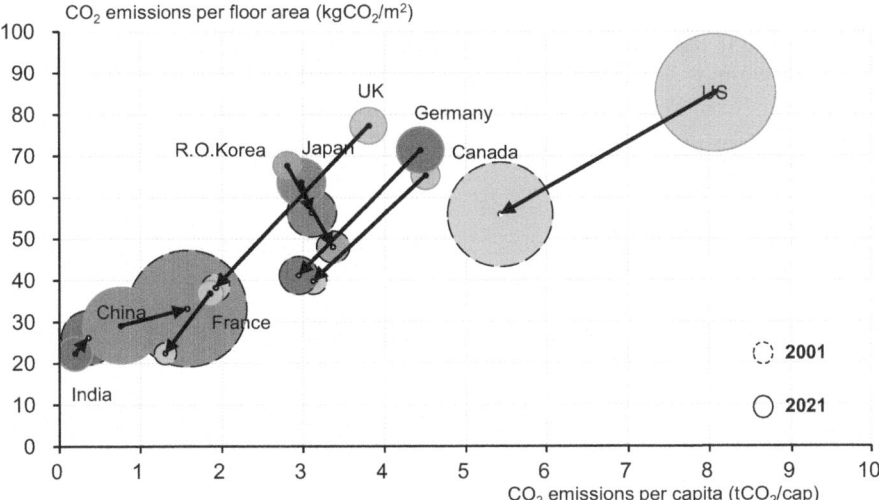

The size of the bubbles represents the total CO_2 emissions of building sector

Fig. 2.7 Comparison of the trends in carbon emissions from building operations between china and foreign countries (2001 and 2021). *Source* Data of countries in 2021 as provided in the CO_2 Emissions from the Fuel Combustion Highlights 2023 database, IEA. Data from China are the results of CBEEM of BERC, Tsinghua University

2001 to 2021. The dashed circles represent carbon emissions data from 2001, while the solid circles represent data from 2021. This figure reflects the changes in per capita carbon emissions and per unit area carbon emissions from building operations in various countries over the past 20 years. Based on these trends, these countries can be classified into three categories. (1) The first category includes countries such as the United States, Canada, Germany, the United Kingdom, and France. A common characteristic among these countries is that the total carbon emissions, per capita emissions, and per unit area emissions from building operations have all shown a downward trend. This is partly due to the reduction in per capita and per unit area energy consumption and also because these countries have actively promoted the transformation of their energy structures, vigorously developing zero-carbon electricity. (2) The second category includes countries like South Korea and Japan. Over the past 20 years, their total carbon emissions have increased, as have their per capita emissions, while per unit area emissions have decreased. The analysis suggests that although the total carbon emissions from building operations in Japan and South Korea have grown slowly in recent years, their population growth rates have been extremely low. Japan's population has even shown negative growth. The growth rate of the population is lower than that of carbon emissions, while the growth rate of building area is higher than that of carbon emissions, leading to inconsistent trends in per capita and per unit area carbon emissions in these two countries. (3) The third category includes countries like China and India. Over the past 20 years, total carbon emissions, per capita emissions, and per unit area emissions have all shown an increasing trend. China and India have both been in a period of rapid development over the past 20 years, with increasing energy intensity. In order to achieve carbon peaking as soon as possible, developing countries such as China and India should simultaneously control the total energy consumption while vigorously promoting the low-carbon transformation of their energy systems.

Open Access This chapter is licensed under the terms of the Creative Commons Attribution-NonCommercial-NoDerivatives 4.0 International License (http://creativecommons.org/licenses/by-nc-nd/4.0/), which permits any noncommercial use, sharing, distribution and reproduction in any medium or format, as long as you give appropriate credit to the original author(s) and the source, provide a link to the Creative Commons license and indicate if you modified the licensed material. You do not have permission under this license to share adapted material derived from this chapter or parts of it.

The images or other third party material in this chapter are included in the chapter's Creative Commons license, unless indicated otherwise in a credit line to the material. If material is not included in the chapter's Creative Commons license and your intended use is not permitted by statutory regulation or exceeds the permitted use, you will need to obtain permission directly from the copyright holder.

Chapter 3
Current Energy Use Situation and Energy-Saving Paths of Rural Buildings

Rural building energy use, as an important component of China's building energy system, has had many problems like primitive ways of energy use, low efficiency, and high pollutant emissions for a long time. Recent years have witnessed great changes in the energy use structure and consumption of rural buildings in China with the support of national action plans including clean space heating in the northern region and winning the blue sky war. By combining data analyses from multiple sources, this chapter presents the current situation of rural building energy consumption in China and the effects of the clean space heating campaign in northern rural areas and analyzes existing problems. On this basis and in combination with the actual demand for heating of northern rural residential buildings (RRBs) in winter, the basic support and key technology development direction for realizing the suitable mode of clean heating in RRBs are proposed, thus providing a reference for the long-term clean heating in northern rural areas and the low-carbon development of rural building energy use in the next step.

3.1 Definitions of Concepts Related to Rural Area

1. Rural Area

The term "rural area" is also commonly called "countryside". According to the provisions in the approval given by the State Council in July 2008 of the *Regulation on Statistical Classification of Urban and Rural Areas* formulated by the National Bureau of Statistics together with the Ministry of Civil Affairs, the Ministry of Housing and Urban–Rural Development, the Ministry of Public Security, the Ministry of Finance, the Ministry of Natural Resources and the Ministry of Agriculture and Rural Affairs (GH [2008] No. 60), the territory of China is classified into urban area and rural area, with the division of administrative areas in China as the foundation, the areas under the jurisdiction of neighborhood committees and villagers' committees as the classification objects and the practical construction as the classification basis. Urban area consists of city areas and counties. The former refers to the areas centering neighborhood committees where municipal government and district governments are located. The latter refers to areas outside of the city, centering neighborhood committees where county-level government is located. Independent industrial and mining areas, development zones, scientific research institutions, colleges and universities, and other special areas where no government is located and the permanent population is over 3,000 as well as the areas where the offices of farms and forest farms are located are considered county areas. Countryside or rural area means an area outside the urban area designated in the abovementioned regulation.

2. Rural Residential Building

Rural residential buildings, abbreviated to RRBs, generally refer to residences that are covered, surrounded by walls, capable of giving protection from wind and rain, and used for people to live and reside in. They are generally residences with a fixed foundation. Depending on the living habits of different places, they include brick houses, stone houses, cave dwellings, bamboo houses, and yurts for people to live and reside in all year round, but exclude houseboats. RRBs are houses primarily used for those engaged in agricultural production to reside in, which are composed of rooms for agricultural production, such as agricultural implements storage places, poultry and livestock breeding places, and other sideline production facilities, as well as general living rooms. Therefore, RRBs are not only a basic living space and an important property for farmers but also a part of the rural means of production.

3. Rural Permanent Population

Rural permanent population means the population in rural areas that are often at home or live at home for more than 6 months in a year and whose financial and living conditions are integrated with their households. Migrant workers who bring most of their income back home and have their financial conditions integrated with their households despite being away from home for more than 6 months are still considered household permanent population. State employees and retirees who live at home and have their living conditions integrated with their households are also considered as household permanent population. However, active-duty soldiers, enrolled students at technical secondary schools and above (other than day students), and migrant workers who are away from home all year round (excluding family visits, seeing a doctor, etc.) and already have a stable career and a place to live are not counted as household permanent population.

4. Rural Building Energy Consumption

Rural building energy consumption refers to the energy consumption generated by the architectural space providing the rural permanent population with the most basic living functions (such as bedroom, living room, kitchen, and bathroom), including the energy consumption generated by heating, cooking, domestic hot water, air conditioning, ventilation, lighting, and home appliances, but excluding that generated by rural production activities and transport activities.

3.2 Current Situation of Rural Building Energy Consumption

At the beginning of China's new countryside construction, Building Energy Research Center (BERC), Tsinghua University, with the support of the Ministry of Agriculture and Rural Affairs and other related institutions, organized over 700 teachers and students to conduct large-scale field survey for the first time in China during the summer of 2006 and 2007 respectively, which covered 24 provinces (autonomous regions and municipalities directly under the Central Government) and a total of 150 county-level administrative regions, in order to fully understand the basic situation of rural building energy use in China. The process of selecting research samples is as follows: First, 24 provinces including northern China and the Yangtze River Basin were selected; next, about 10 counties (cities) were randomly selected from each province; then, 5–6 villages were randomly selected from each county (city), and 5–6 households were randomly selected from each village. Based on the detailed research data on these sample households and in combination with the comprehensive data of the rural energy offices of the villages, towns, counties, and provinces, first-hand data including the rural residential building conditions, different energy use categories, and energy consumption of the provinces were obtained by "bottom-up" counting and "top-down" verification. The detailed research results and analysis of this part are provided in the *2012 Annual Report on China Building Energy Efficiency*.

Since 2005, with the promotion of new countryside construction and the improvement of farmers' living standards, as well as the implementation of national strategies of building a moderately prosperous society in all respects, building a beautiful countryside, etc., rural residential buildings and energy consumption have had new impacts. To further understand the changes in the energy consumption of rural residential buildings in different regions of China in nearly a decade after 2005 and the impact of rural energy use on outdoor atmospheric environmental pollution that had become an issue of general social concern, BERC, Tsinghua University, in collaboration with Beijing Sustainable Development Promotion Association, organized large-scale comprehensive survey on China's rural energy environment again in the summer of 2015, with the support of the Ministry of Science and Technology and Beijing Municipal Science and Technology Commission. A total of 21 research teams and over 200 teachers and students participated in the research after receiving unified training. The research methods and organization were similar to those of the large-scale national research in 2006 and 2007. The method of sampling at provincial, county, and village levels was still adopted, and the research covered 11 provinces (autonomous regions and municipalities directly under the Central Government) of northern China, i.e., Beijing, Tianjin, Hebei, Shandong, Shaanxi, Heilongjiang, Liaoning, Inner Mongolia, Gansu, Ningxia and Qinghai, and 10 provinces (autonomous regions and municipalities directly under the Central Government) of southern China, i.e., Zhejiang, Jiangsu, Anhui, Jiangxi, Hunan, Chongqing, Sichuan, Fujian, Yunnan, and Guizhou. The survey mainly comprised the current consumption and specific proportions of rural residential building energy; the current situation and development trend of rural renewable energy utilization; the conditions of rural residential building and household energy consumption (including house heating, cooking, and cooling by air conditioner); the conditions of ecological environment and comprehensive resource utilization; and the current situation and demand of the development of economy, technical information, and industry, etc., in rural areas. The detailed survey results and analysis of this part are provided in the *2016 Annual Report on China Building Energy Efficiency*.

With the continual advancement of clean space heating in northern China in recent years, the state has successively issued multiple clean space heating policies, which have had a significant impact on both the total amount and structure of rural energy consumption. For example, the *Clean Winter Heating Plan in Northern China (2017–2021)* released by ten ministries and commissions including the National Development and Reform Commission states that by 2021 the overall clean space heating rate in northern China should reach 70% and 150 million tons of bulk coal (including coal for inefficient small boilers) should be replaced. Specifically, for the "2+26" key cities, all urban areas should achieve clean heating and dismantle all coal-fired boilers with steam output below 35 tons per hour; counties and rural–urban fringes should reach a clean heating rate of more than 80% and dismantle all coal-fired boilers with steam output below 20 tons per hour; and rural areas should reach a clean heating rate of more than 60%.

To get an in-depth knowledge of the current situation of rural building energy consumption in China, the energy consumption intensity of provinces obtained from the research samples is taken as a benchmark, and such parameters as rural population and households of provinces (autonomous regions and municipalities directly under the Central Government) provided in the *China Statistical Yearbook 2023* are used to calculate the total energy consumption of rural buildings in 30 provinces (autonomous regions and municipalities directly under the Central Government) in 2022. The obtained result is about 253 million tce, including the energy consumption of heating, cooking (including domestic hot water), air conditioning, and household electricity consumption (including lighting and various home appliances), and the energy types counted include commercial energy such as coal (bulk coal and honeycomb briquette), liquefied petroleum gas (LPG), electricity and natural gas and non-commercial energy dominated by firewood and straw. Among them, electricity is converted into kilograms of coal equivalent (kgce) according to the calculation method of coal consumption of thermal power supply in the current year, and other types of energy are converted based on the average net calorific values of fuels[1]. As shown in Table 3.1, in the energy consumption of rural buildings, the consumption of commercial energy coal, LPG, electricity, and natural gas is 95 million tons (67 million tce), 8.07 million tons (14 million tce), 348.1 billion kWh (105 million tce) and 12.4 billion m^3 (15 million tce) respectively, totaling 201 million tce; the total consumption of non-commercial energy biomass (including firewood and straw) is 94 million tons (52 million tce); and commercial energy and non-commercial energy account for 79.4% and 20.6% respectively.

---

[1] 1 kWh = 0.301 kgce; 1 kg coal = 0.71 kgce; 1 kg LPG = 1.71 kgce; 1 kg firewood = 0.6 kgce; 1 kg straw = 0.5 kgce; 1 m^3 natural gas = 1.21 kgce.

3.2 Current Situation of Rural Building Energy Consumption

Table 3.1 Consumption of different types of energy in rural household energy consumption in some provinces (autonomous regions and municipalities directly under the Central Government) of China in 2022

Province (autonomous region and municipality directly under the Central Government)	Rural permanent population (10,000 people)	Annual physical consumption						Coal equivalent (10,000 tce)		
		Coal (10,000 t)	Liquefied gas (10,000 t)	Electric energy (100 million kWh)	Firewood (10,000 t)	Straw (10,000 t)	Natural gas (100 million m³)	Commercial energy	Non-commercial energy	Total
Beijing	271.5	29	8.4	67.8	1.2	0.8	3.2	277	1	279
Tianjin	203	10	3.7	37.5	0.1	1.2	9	235	1	236
Hebei	2845	192	11.5	575.5	4.2	74.6	38	2348	40	2388
Shanxi	1254.9	380	1.9	277	9.3	127	14.6	1283	69	1352
Inner Mongolia	754	929	25.3	62.7	13.7	111	2.3	920	64	983
Liaoning	1133	668	24.5	83.5	277	1040	2.8	802	686	1488
Jilin	851.5	331	4.0	18.9	184	473	1.2	313	347	660
Heilongjiang	1047	964	8.0	32.9	1413	877	1.8	819	1286	2105
Shanghai	205	/	12.0	14.8	2.4	4.8	0.2	67	4	71
Jiangsu	2180	274	71.3	305	227	43.4	1.4	1252	158	1409
Zhejiang	1749	228	62.9	139.5	193	122	1.2	704	177	881
Anhui	2441	239	70.7	120.5	847	92.9	0.8	663	554	1218
Fujian	1251	34	29.8	136.9	463	0.9	0.8	497	278	775
Jiangxi	1717.5	279	37.2	80.1	526	85.3	0.5	508	358	867
Shandong	3603.7	184	45.7	254	213.6	89.5	18.7	1200	173	1373
Henan	4239	312	21.8	182.1	36.3	31.3	2.5	837	37	875
Hubei	2065	292	14.1	50.7	44.4	19.1	0.4	389	36	425

(continued)

Table 3.1 (continued)

Province (autonomous region and municipality directly under the Central Government)	Rural permanent population (10,000 people)	Annual physical consumption						Coal equivalent (10,000 tce)		
		Coal	Liquefied gas	Electric energy	Firewood	Straw	Natural gas	Commercial energy	Non-commercial energy	Total
		(10,000 t)	(10,000 t)	(100 million kWh)	(10,000 t)	(10,000 t)	(100 million m^3)			
Hunan	2621	447	81.4	110.1	16	44.9	0.6	796	32	828
Guangdong	3191.4	172	69.4	91.5	29.4	148	2.1	542	91	633
Guangxi	2238	220	2.1	100.8	13.6	12.4	0.4	468	14	482
Hainan	395.5	38	1	10	3.3	11.1	0.4	64	8	71
Sichuan-Chongqing	4420.8	435	169	265.2	117	508	2.9	1431	324	1755
Guizhou	1742	834	9.3	83.1	42.9	31.1	0.5	864	41	906
Yunnan	2266	13	3.3	59.1	24.4	194	0.8	203	112	315
Xizang	228	/	/	/	26.5	/	/	/	16	16
Shaanxi	1424	177	12.2	230	13.4	31.4	10.5	966	24	989
Gansu	1141.8	368	3.6	46.3	18.6	33	2.2	433	28	461
Qinghai	229	123	1.1	12.7	149	58.4	1.6	147	119	265
Ningxia	245	64	0.5	13.1	127	76	0.9	97	114	211
Xinjiang	1089	1249	1.1	20	18	23.7	1.2	963	23	986
Northern region	20559	5980	173	1914	2479	3047	111	11640	3011	14651

(continued)

3.2 Current Situation of Rural Building Energy Consumption

Table 3.1 (continued)

Province (autonomous region and municipality directly under the Central Government)	Rural permanent population (10,000 people)	Annual physical consumption						Coal equivalent (10,000 tce)		
		Coal (10,000 t)	Liquefied gas (10,000 t)	Electric energy (100 million kWh)	Firewood (10,000 t)	Straw (10,000 t)	Natural gas (100 million m^3)	Commercial energy	Non-commercial energy	Total
Southern region	28483	3505	633	1567	2548	1319	13	8448	2188	10636
Total	49042	9485	807	3481	5053	4365	124	20088	5215	25303

Note The data of Sichuan-Chongqing in the table includes Sichuan Province and Chongqing Municipality; Hong Kong, Macao, and Taiwan regions are not counted in this table due to fewer traditional rural areas (hereinafter the same)

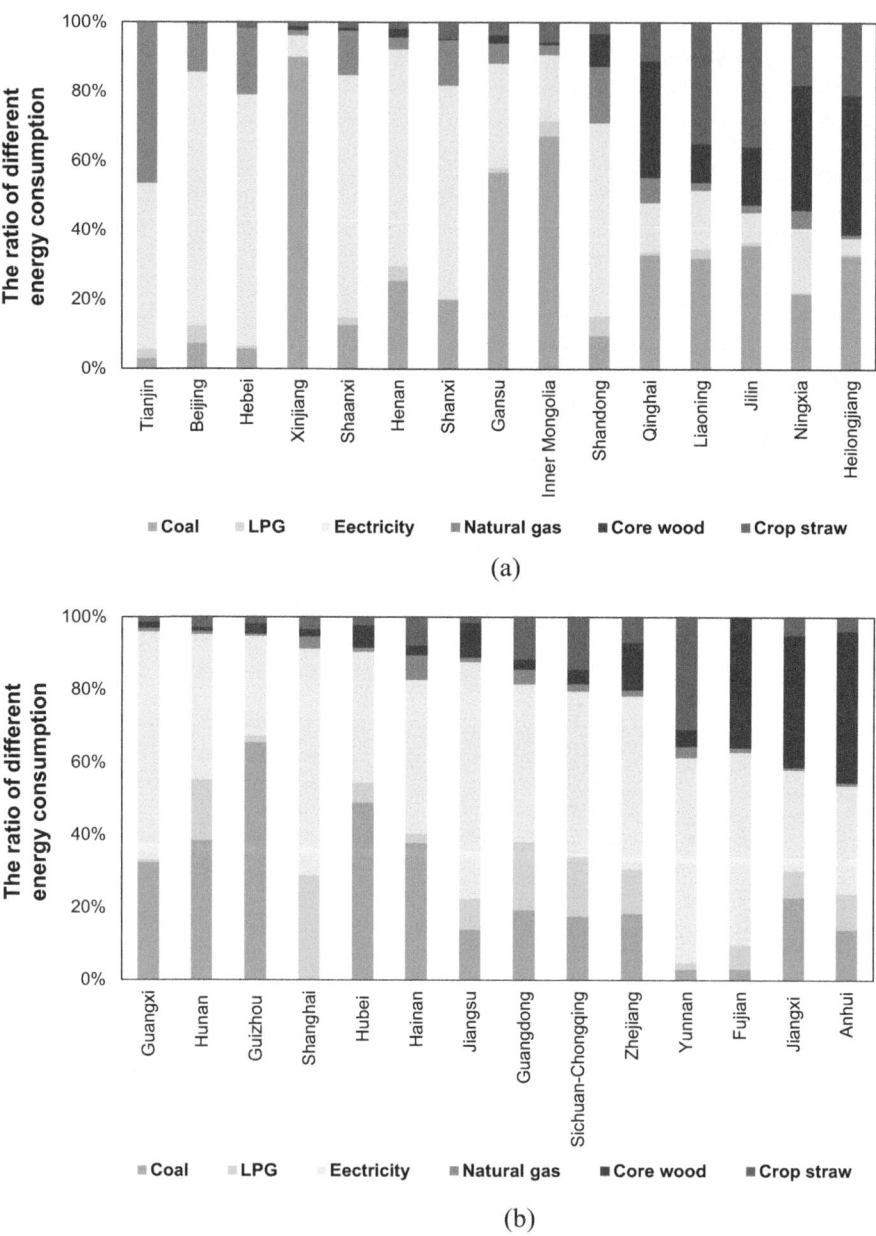

Fig. 3.1 Distribution of proportions of different energy consumption of rural buildings in some provinces (autonomous regions and municipalities directly under the Central Government) of China. **a** Northern region; **b** Southern region

Figure 3.1a and b presents the distribution of proportions of different types of energy consumption of rural buildings in some provinces (autonomous regions and municipalities directly under the Central Government) in the northern and southern regions of China throughout the year respectively. As shown in the figures, the commercial energy consumption of the 9 provinces (autonomous regions and municipalities directly under the Central Government) in the northern region, i.e., Tianjin, Beijing, Hebei, Xinjiang, Shaanxi, Henan, Shanxi, Gansu, and Inner Mongolia, accounts for more than 90%, and only in Jilin, Ningxia, and Heilongjiang, the commercial energy consumption accounts for less than 50% and the consumption of non-commercial energy including firewood and straw accounts for a high proportion; the commercial energy consumption of 5 provinces in the southern region, i.e., Guangxi, Hunan, Guizhou, Shanghai, and Hubei, accounts for more than 90%, and only in Jiangxi and Anhui, the commercial energy consumption accounts for less than 60% and the consumption of non-commercial energy including firewood and straw accounts for a relatively high proportion.

Over the past few decades, bulk coal, as an important energy source in rural areas of China, has been widely distributed in various places.

However, it has become a major pollution source due to its low combustion efficiency, a lack of terminal flue gas cleaning devices, and the difficulty in emission regulation. In recent years, governments at all levels in China have carried out large-scale bulk coal reduction actions and achieved certain results, which have significantly impacted the distribution of bulk coal in various regions. To see the current distribution of bulk coal in various regions and provinces in China more intuitively and thus find the governance priorities for the next step, this chapter divides the 31 provinces, municipalities directly under the Central Government and autonomous regions of China, other than Hong Kong, Macao and Taiwan, into 4 regions, namely "Beijing-Tianjin-Hebei" and surrounding areas (including 7 provincial-level administrative regions, i.e., Beijing, Tianjin, Hebei, Shandong, Shanxi, Shaanxi, and Henan), other northern provinces except for "Beijing-Tianjin-Hebei" and surrounding areas (including 9 provincial-level administrative regions, i.e., Heilongjiang, Jilin, Liaoning, Inner Mongolia, Gansu, Ningxia, Qinghai, Xinjiang, and Xizang), provinces in the Yangtze River Basin (including 9 provincial-level administrative regions, i.e., Shanghai, Anhui, Jiangsu, Zhejiang, Hubei, Hunan, Jiangxi, Sichuan, and Chongqing) and other southern provinces except for those in the Yangtze River Basin (including 6 provincial-level administrative regions, i.e., Guizhou, Yunnan, Guangdong, Guangxi, Fujian, and Hainan). Figure 3.2a–d presents the per capita annual physical consumption of bulk coal in rural buildings in these 4 regions respectively. The bulk coal consumption in rural areas in Shanghai and Xizang is negligible and thus is not shown in the figures.

The per capita bulk coal consumption of local rural buildings is calculated based on the total population of each region in 2022. The obtained per capita bulk coal consumption in other northern provinces except for "Beijing-Tianjin-Hebei" and surrounding areas is the highest, reaching 724 kg/(capita·a), mainly because most of these provinces are located in severe cold areas with high energy consumption for winter heating, the number of cities in each province that have been shortlisted for the National Northern Clean Heating Pilot Project is generally not more than 5, such

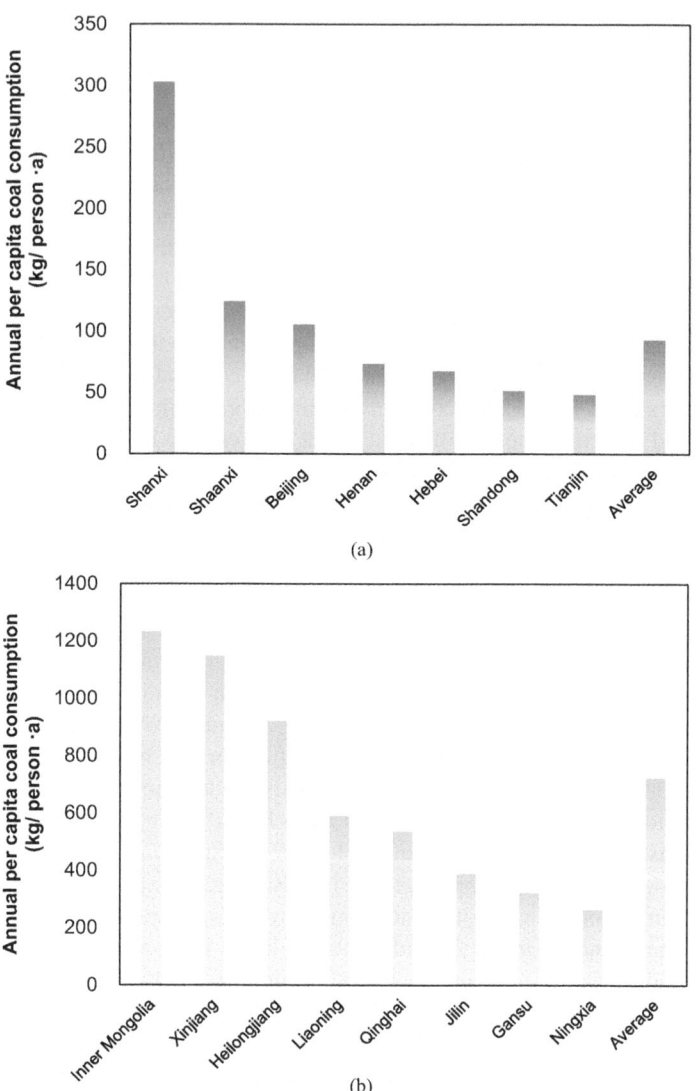

Fig. 3.2 Per capita annual physical consumption of bulk coal in rural buildings in different regions of China (I). **a** "Beijing-Tianjin-Hebei" and surrounding areas; **b** Other northern provinces except for "Beijing-Tianjin-Hebei" and surrounding areas. Per capita annual physical consumption of bulk coal in rural buildings in different regions of China (II). **c** Provinces in the Yangtze River Basin; **d** Other southern provinces except for those in the Yangtze River Basin

3.2 Current Situation of Rural Building Energy Consumption

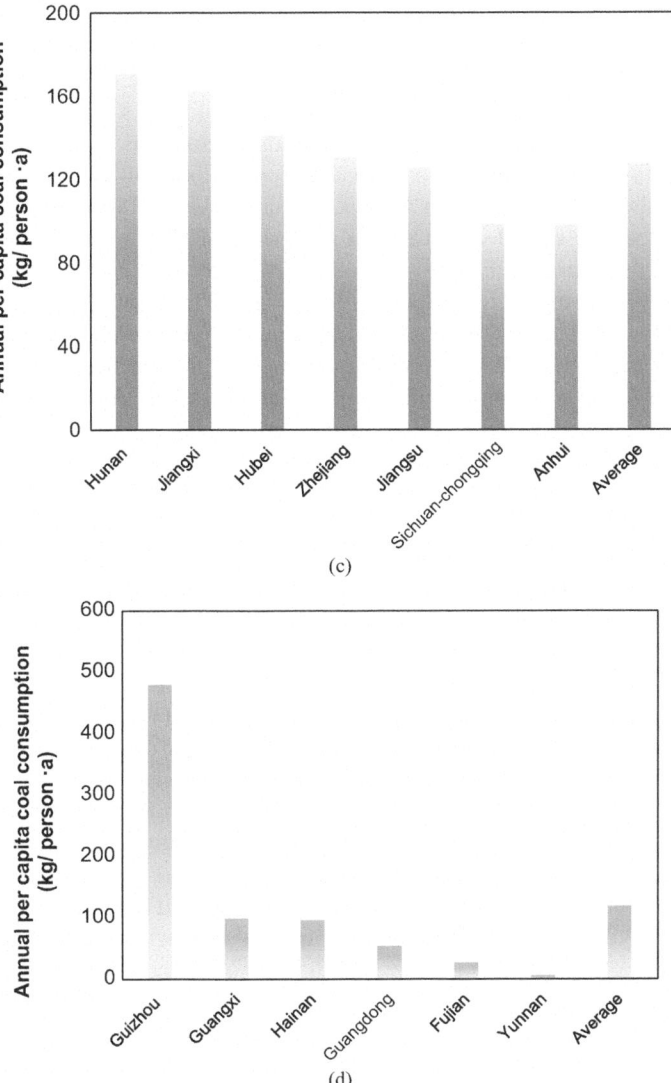

Fig. 3.2 (continued)

projects have just been carried out for one to two years and is in the initial stage, and not many farmer households have completed building energy efficiency retrofitting and clean energy substitution. In this region, the per capita bulk coal consumption in Inner Mongolia is the highest, reaching 1,233 kg/(capita·a). The per capita coal consumption of rural buildings in provinces in the Yangtze River Basin is 128 kg/(capita·a), mainly because this region is in a hot summer and cold winter (HSCW) climate zone where a significant number of farmers need coal for heating or providing cooking energy in winter and there is no clean heating campaign similar to that in the northern region to replace bulk coal, causing the coal consumption intensity to remain almost unchanged from previous years. There is little difference in the total coal consumption of these provinces, all in the order of three to four million tons. In this region, the per capita bulk coal consumption in Hunan is the highest, reaching 171 kg/(capita·a). The per capita coal consumption in other southern provinces except for those in the Yangtze River Basin ranks third, reaching 118 kg/(capita·a). In this region, the total bulk coal consumption and per capita bulk coal consumption in Guizhou are much higher than those in other provinces, mainly because Guizhou is one of the few coal-producing provinces in southern China and located on a plateau where the climate is cold in winter, causing many farmer households to use coal for heating. The per capita bulk coal consumption in Guizhou reaches 479 kg/(capita·a). The per capita coal consumption in "Beijing-Tianjin-Hebei" and surrounding areas is the lowest, only 93 kg/(capita·a), which is mainly due to the clean heating pilot campaign that has been carried out by the state on a large scale in this region since 2017. By 2021, all cities had been shortlisted for the pilot program. At present, as some cities are still in the ending stage of bulk coal substitution, bulk coal has not been completely cleared. In this region, Shanxi has the highest per capita bulk coal consumption, reaching 303 kg/(capita·a), while Tianjin has the lowest per capita bulk coal consumption, only 48 kg/(capita·a).

Currently, among the farmer households in the northern region that do not implement clean space heating transformation, heating is still the mainstay of energy consumption and bulk coal is the main energy form. Take a village in Inner Mongolia as an example. This village has a total area of 13.25 km^2 and a total of 240 households, including 188 permanent households. It has a total arable area of 1,757.2 μ (1.1715 km^2) and an average household arable area of 7 μ (4,666.7 m^2), with corn, potato, and sorghum as the main crops. Straw resources are mainly used for mulching and only a small amount is used as fuel. Residential buildings are concentrated, about 1 km from surrounding villages, and the average household floor area is 120 m^2. The permanent population is mainly the elderly and children, with an average household permanent population of 2–3 people, and the economic income hinges on planting and breeding. The village has a temperate and continental monsoon climate. It is arid, windy, and cold, with plenty of daylight and a large temperature difference. The winter is long and severely cold, with an average temperature of minus 13 °C in January and a heating period of 150 days.

From this village, 90 households were selected for detailed research. The results reveal that more than 90% of the rural houses are the large three-room type, as shown in Fig. 3.3. Most of the rural houses have been renovated or newly built since 2000, and a few are old houses from the twentieth century. Rural houses are all masonry-concrete structure, 90% of them have 37 cm thick solid brick walls, and a few have 50 cm thick adobe walls. Over 85% of the houses have aluminum alloy double-glass windows for doors and windows, a few have double windows, and less than 2% of the houses have sunrooms, as shown in Fig. 3.4. Rural houses have poor thermal insulation properties, and their roofs are all thermal insulation structures made of sunflower stalks and plastering. More than 90% of the houses

3.2 Current Situation of Rural Building Energy Consumption

Fig. 3.3 Typical rural houses in a village in Inner Mongolia

Fig. 3.4 Sunrooms and double-window rural houses in a village in Inner Mongolia

have no thermal insulation for exterior walls, and a few have exterior walls covered with 5 cm thick polystyrene boards.

Farmer households in this village have coal-fired boilers + radiators and scattered burning of biomass + heatable brick bed as the main heating methods, which are also used for cooking in winter, as shown in Fig. 3.5. The actual average household heating area is about 60 m^2. The average household bulk coal consumption is 3–5 t/year, accounting for more than 80% of the annual total energy consumption of households. The annual heating fee is about RMB 4,000. The average indoor temperature in winter is 14 °C.

Fig. 3.5 Winter heating facilities for farmer households in a village

The *2008 Annual Report on China Building Energy Efficiency* systematically provides the overall energy consumption and energy consumption by province of rural buildings in China in 2006 for the first time. The consumption of commercial energy coal, LPG, and electricity was 190 million tons (136 million tce), 5.97 million tons (10 million tce), and 132.4 billion kWh (44 million tce) respectively. The total consumption of non-commercial energy biomass (including firewood and straw) was 220 million tons (124 million tce)[2]. Commercial energy and non-commercial energy accounted for 61% and 39%, respectively. After 16 years of development and changes, commercial energy and non-commercial energy accounted for 79.4% and 20.6% respectively in 2022, and the proportion of non-commercial energy dropped by nearly 20% compared to 2006. This change in energy structure has played a positive role in controlling air pollutant emissions but has also led to some issues requiring special attention. This change does not happen naturally without guidance but is closely related to the clean heating campaign in northern China.

3.3 Clean Winter Heating Campaign in Northern China and Its Effects

At the 14th Meeting of the Central Leading Group for Financial and Economic Affairs at the end of 2016, General Secretary Xi Jinping stressed that promoting clean heating in northern China during winter has a bearing on the ability of the broad masses in northern China to keep warm and on whether the haze weather can be reduced and also forms an essential part of energy production and consumption revolution as well as changes in the rural lifestyle. He also stated that, based on the policy that enterprises assume the main responsibility and governments provide support to ensure affordable heating for the people, the most appropriate form of energy should be selected, clean energy should be utilized whenever possible, and the increase in the proportion of clean heating should be accelerated. Specifically, clean heating means a method of heating with electricity, terrestrial heat, biomass, solar energy, industrial waste heat, clean coal (ultra-low emission), natural gas, nuclear energy, and other clean energy to realize low emission and low energy consumption via an efficient energy use system, which includes the whole process of heating aiming at reducing pollutant emission and energy consumption and involves such links as clean heat, efficient transmission and distribution network (heat network), and energy-efficient buildings (heat user). Promoting clean heating is of great significance to the reduction of energy consumption for heating, the improvement of energy efficiency, and the winning of the blue-sky war and the critical battle against pollution.

To fully implement the important instructions of General Secretary Xi Jinping on promoting clean heating in northern China in winter, ten ministries and commissions including the National Development and Reform Commission, the Ministry of Environmental Protection, and the National Energy Administration jointly issued

---

[2] At that time, 1 kWh = 0.333 kgce.

3.3 Clean Winter Heating Campaign in Northern China and Its Effects

the *Clean Winter Heating Plan in Northern China (2017–2021)* in 2017 to comprehensively promote clean heating in northern China. The Plan states that: By 2021, the clean heating rate in northern China should reach 70%, 150 million tons of bulk coal (including coal for inefficient small boilers) should be replaced, the average comprehensive energy consumption of heating systems should be reduced to below 15 kgce/m^2, and the proportion of existing energy-efficient residential buildings in northern urban areas should reach 80%. The goal is to basically realize clean bulk coal heating in urban areas with severe haze and form a fair and open clean heating market with diversified operations and a high level of service in about 5 years.

In June 2017, the four ministries and commissions, i.e., the Ministry of Finance, the Ministry of Housing and Urban–Rural Development, the Ministry of Environmental Protection, and the National Energy Administration, jointly released the *Notice on Carrying Out the Pilot Work on Clean Winter Heating in Northern China Supported by Central Budget* (CJ [2017] No. 238), which clarified the pilot work on clean winter heating in northern China supported by the central budget. The pilot work focused on supporting the "2+26" cities[3], i.e., air pollution transmission channels in Beijing-Tianjin-Hebei and surrounding areas, and the first batch of 12 pilot cities were determined through competitive reviews. During the three-year pilot demonstration period, each municipality directly under the Central Government was allocated RMB 1 billion per year, each provincial capital was allocated RMB 0.7 billion per year, and each prefecture-level city was allocated RMB 0.5 billion per year.

In August 2018, the Ministry of Finance, the Ministry of Ecology and Environment, the Ministry of Housing and Urban–Rural Development, and the National Energy Administration jointly issued the *Notice on Expanding Pilot Cities for Clean Winter Heating in Northern China Supported by Central Budget* (CJ [2018] No. 397). The pilot cities were expanded to air pollution transmission channel cities in Beijing-Tianjin-Hebei and surrounding areas, Zhangjiakou City, and cities on the Fenwei Plain.

In July 2019, the four ministries and commissions jointly issued the *Notice on Releasing Budget for Air Pollution Prevention and Control in 2019* (CZH [2019] No. 6), which clarified the list of the third batch of pilot cities for clean heating and the amount of fund subsidies.

On April 27, 2022, the Bureau of Natural Resources, Ecology and Environment under the Ministry of Finance, the Department of Standards and Norms under the Ministry of Housing and Urban–Rural Development, the Department of Atmospheric Environment under the Ministry of Ecology and Environment, and the General Affairs Department under the National Energy Administration jointly issued the *Announcement of the List of Clean Winter Heating Projects in Northern China to Be Supported in 2022*, which identified 25 projects as the clean winter heating projects

---

[3] It includes Beijing, Tianjin, Shijiazhuang, Tangshan, Handan, Xingtai, Baoding, Cangzhou, Langfang, and Hengshui in Hebei Province, Taiyuan, Yangquan, Changzhi, and Jincheng in Shanxi Province, Jinan, Zibo, Jining, Dezhou, Liaocheng, Binzhou, and Heze in Shandong Province, Zhengzhou, Kaifeng, Anyang, Hebi, Xinxiang, Jiaozuo, and Puyang in Henan Province, or "2+26" cities for short.

in northern China to be supported by the air pollution prevention and control fund in 2022. After the addition of the first four batches of 63 cities that have been included in the pilot work on clean winter heating supported by the central budget during 2017–2021, the total number of cities and projects supported reached 88. The pilot cities and projects for clean heating in northern China during 2017–2022 are shown in Table 3.2.

3.3.1 Energy Conservation and Emission Reduction Effects

China launched the *Air Pollution Prevention and Control Action Plan*, the first action plan for the comprehensive governance of prominent environmental issues, in September 2013. Beijing and other representative northern cities were the first to start comprehensive haze governance, which was then gradually expanded to other cities. Therefore, the year 2014 can be considered the starting year and the base year of rural atmospheric environmental governance. In addition, BERC, Tsinghua University once organized a large-scale investigation on the energy and environmental conditions in rural areas of China in 2014. By comparing the energy use data of rural buildings in 2022 and 2014, the effects of the implementation of major actions including clean heating in northern China in the past nearly 10 years can be found intuitively.

According to the results provided in the *2023 Report on Development of China's Clean Heating Industry*, by the end of 2022, northern China has reached a clean heating area of 17.9 billion m² and a clean heating rate of 75%, indicating the achievement of the set target of reaching a clean heating rate of 70% and replacing 150 million tons of bulk coal, with a substitution rate of 75%, in which rural areas contribute to 102 million tons of bulk coal substitution. In each pilot city and project for clean heating in northern China promoted in the past, the transformation implementation schemes all included urban and rural areas. However, for most regions, rural areas are the focus and difficulty of the work. The following data will focus on the analysis of the results of clean heating in rural areas.

Table 3.3 provides a comparison of the consumption of different types of energy in the energy use of rural buildings in some provinces and cities of China in 2022 and 2014, from which we can see the combined influence of factors including the change in rural population, the change in the way of energy use, the improvement of energy use efficiency and the clean heating campaign.

The total energy consumption of rural buildings in China dropped by 22.6% from 327 million tce in 2014 to 253 million tce in 2022 as a result of the combined action of multiple factors, including the rising per capita income level in rural areas and certain changes in the energy consumption capacity and concept of farmer households. A slight increase in the per capita energy consumption intensity may be caused; however, driven by the urbanization process from 2014 to 2022, the rural population in China decreased by 17.8%, which eventually led to a reduction of the total energy consumption. Moreover, during this period, the state continually promoted

3.3 Clean Winter Heating Campaign in Northern China and Its Effects

building energy efficiency retrofitting and clean energy substitution. Rural building energy efficiency retrofitting causes the demand for energy consumption for heating to reduce; furthermore, both the approach of "switching from coal to heat pumps"

Table 3.2 Pilot cities and projects for clean heating in northern China during 2017–2022

Pilot batch	Publicity time	Pilot city	Central subsidy funds
First batch	June 2017	[Municipality directly under the Central Government] Tianjin	RMB 1 billion/a each
		[Provincial capital] Shijiazhuang, Taiyuan, Jinan, and Zhengzhou	RMB 0.7 billion/a each
		[Prefecture-level city] Tangshan, Baoding, Langfang, Hengshui, Kaifeng, Hebi and Xinxiang	5/a RMB 0.5 billion/a each
Second batch	August 2018	226 [2+26 cities] Handan, Xingtai, Zhangjiakou, Cangzhou, Yangquan, Changzhi, Jincheng, Zibo, Jining, Binzhou, Dezhou, Liaocheng, Heze, Anyang, Jiaozuo and Puyang	RMB 0.5 billion/a each
		[City on the Fenwei Plain] Lvliang, Jinzhong, Linfen, Yuncheng, Luoyang, Xian, and Xianyang	RMB 0.3 billion/a each

(continued)

Table 3.2 (continued)

Pilot batch	Publicity time	Pilot city	Central subsidy funds
Third batch	July 2019	[City on the Fenwei Plain] Tongchuan, Weinan, Baoji and Sanmenxia	RMB 0.3 billion/a each
		[Other cities] Dingzhou, Xinji, Jiyuan, and Yangling Demonstration Zone	RMB 0.1 billion/a each
Fourth batch	April 2021	[Key area] Beijing, Chengde, Qinhuangdao, Xuchang, Yantai, Weifang, Taian, Datong, Xinzhou, Shuozhou, Yanan and Yulin	RMB 1 billion/a to Beijing and RMB 0.3 billion/a to each of the other cities
		[Other areas] Fuxin, Jiamusi, Baotou, Haixi, Urumqi, Liaoyuan, Lanzhou and Wuzhong	
Fifth batch	April 2022	[Provincial capital] Hohhot, Xining, Yinchuan, Harbin, Changchun, and Shenyang	RMB 0.7 billion/a each

(continued)

3.3 Clean Winter Heating Campaign in Northern China and Its Effects

Table 3.2 (continued)

Pilot batch	Publicity time	Pilot city	Central subsidy funds
		[Other cities] Ulanqab, Bayannur, Zhongwei, Guyuan, Wuwei, Jinchang, Linxia Hui Autonomous Prefecture, Changji Hui Autonomous Prefecture, Xinjiang Production and Construction Corps (7th, 8th, and 13th Divisions), Qiqihar, Jilin, Baishan, Panjin, Yingkou, Zhoukou, Shangqiu, Qingdao, Zaozhuang and Dongying	RMB 0.3 billion/a each
Total	88 cities		RMB 107.1 billion

and the approach of "switching from coal to gas" have much higher energy efficiency than traditional bulk coal furnaces in rural areas and can obtain more net heat when consuming the same amount of energy, which has also contributed to a further reduction of the total energy consumption.

In terms of individual energy consumption, the total coal consumption declined obviously by about 51.8% from 197 million tons in 2014 to 95 million tons in 2022; natural gas consumption, which was very small and negligible in 2014, had amounted to a net increase of 12.4 billion m^3 by 2022; LPG remained roughly flat from 8.31 million tons in 2014 to 8.07 million tons in 2022; electric energy increased from 214 billion kWh in 2014 to 348.1 billion kWh in 2022; total biomass consumption further decreased by over 87 million tons from 181 million tons in 2014 to 94 million tons in 2022, and its proportion declined rapidly from 31.2% to 20.6%.

The direct burning of bulk coal, straw, firewood, and other solid fuels by households in rural areas is recognized as one of the main causes of environmental pollution and regional and global climate change and is also a major environmental health risk factor. The direct burning of solid fuels in rural areas will produce a large number

Table 3.3 Comparison of total consumption of different types of energy in the energy use of rural buildings in some provinces (autonomous regions and municipalities directly under the Central Government) of China in 2022 and 2014

Province (autonomous region and municipality directly under the Central Government)	Rural permanent population (10,000 people)		Annual physical consumption											
			Coal (10,000 t)		Liquefied gas (10,000 t)		Electric energy (100 million kWh)		Fire wood (10,000 t)		Straw (10,000 t)		Natural gas (100 million m^3)	
	2022	2014	2022	2014	2022	2014	2022	2014	2022	2014	2022	2014	2022	2014
Beijing	271.5	294	29	571	8.4	18.4	67.8	56.4	1.2	96	0.8	63	3.2	–
Tianjin	203	269	10	196	3.7	7.7	37.5	12.7	0.1	3	1.2	54	9	–
Hebei	2845	3741	192	1631	11.5	25.6	575.5	92.7	4.2	34	74.6	610	38	–
Shanxi	1254.9	1686	380	2609	1.9	2.6	277	19.5	9.3	72	127	878.9	14.6	–
Inner Mongolia	754	1014	929	1226	25.3	27.9	62.7	44.5	13.7	18	111	146	2.3	–
Liaoning	1133	1447	668	797	24.5	27	83.5	66.8	277	331	1040	1240	2.8	–
Jilin	851.5	1244	331	388	4.0	4.8	18.9	14.8	184	216	473	555	1.2	–
Heilongjiang	1047	1609	964	1406	8.0	13.3	32.9	34.4	1413	2059	877	1278	1.8	–
Shanghai	205	252	–	–	12.0	12.4	14.8	14.9	2.4	3	4.8	6	0.2	–
Jiangsu	2180	2769	274	354	71.3	76.6	305	320.4	227	293	43.4	56	1.4	–
Zhejiang	1749	1935	228	224	62.9	51.4	139.5	111.2	193	189	122	120	1.2	–
Anhui	2441	3093	239	219	70.7	53.9	120.5	89.7	847	775	92.9	85	0.8	–
Fujian	1251	1454	34	38	29.8	28	136.9	125.8	463	523	0.9	1	0.8	–
Jiangxi	1717.5	2261	279	343	37.2	38.2	80.1	80.2	526	648	85.3	105	0.5	–
Shandong	3603.7	4404	184	905	45.7	49.2	254	170.4	213.6	1019	89.5	427	18.7	–
Henan	4239	5171	312	1846	21.8	20.9	182.1	76.3	36.3	205	31.3	185	2.5	–
Hubei	2065	2578	292	442	14.1	17.4	50.7	61.4	44.4	427	19.1	29	0.4	–

(continued)

3.3 Clean Winter Heating Campaign in Northern China and Its Effects

Table 3.3 (continued)

Province (autonomous region and municipality directly under the Central Government)	Rural permanent population (10,000 people)		Annual physical consumption										
			Coal (10,000 t)		Liquefied gas (10,000 t)		Electric energy (100 million kWh)		Fire wood (10,000 t)		Straw (10,000 t)		Natural gas (100 million m³)	
	2022	2014	2022	2014	2022	2014	2022	2014	2022	2014	2022	2014	2022	2014
Hunan	2621	3417	447	588	81.4	99.1	110.1	130.8	16	505	44.9	59	0.6	–
Guangdong	3191.4	3432	172	249	69.4	79.6	91.5	102.5	29.4	445	148	214	2.1	–
Guangxi	2238	2567	220	159	2.1	1.4	100.8	67.1	13.6	287	12.4	9	0.4	–
Hainan	395.5	418	38	41	1	0.8	10	7.5	3.3	25	11.1	12	0.4	–
Sichuan-Chongqing	4420.8	5580	435	410	169	139.8	265.2	214.5	117	1302	508	479	2.9	–
Guizhou	1742	2104	834	752	9.3	6.8	83.1	59.4	42.9	344	31.1	28	0.5	–
Yunnan	2266	2747	13	15	3.3	3.4	59.1	59	24.4	710	194	223	0.8	–
Xizang	228	236	–	–	–	–	–	4.8	26.5	202	–	–	–	–
Qinghai	229	293	123	187	1.1	1	12.7	6.6	149	227	58.4	89	1.6	–
Ningxia	245	307	64	145	0.5	0.6	13.1	9.2	127	286	76	171	0.9	–
Xinjiang	1089	1240	1249	1108	1.1	0.8	20	8.8	18	16	23.7	21	1.2	–
Northern region	20559	26257	5980	15842	173	222	1914	696	2479	4945	3047	6143	111	–
Southern region	28483	34607	3505	3834	633	609	1567	1445	2548	6476	1319	1426	13	–
Total	49042	60864	9485	19676	807	831	3481	2140	5053	11421	4365	7569	124	–

(continued)

Table 3.3 (continued)

Province (autonomous region and municipality directly under the Central Government)	Coal equivalent (10,000 tce)						
	Commercial energy		Non-commercial energy		Total		
	2022	2014	2022	2014	2022	2014	
Beijing	277	612	1	88	279	700	
Tianjin	235	191	1	30	236	221	
Hebei	2348	1481	40	342	2388	1823	
Shanxi	1283	1925	69	483	1352	1983	
Inner Mongolia	920	1068	64	181	983	1249	
Liaoning	802	833	686	819	1488	1652	
Jilin	313	335	347	407	660	742	
Heilongjiang	819	1139	1286	1875	2105	3014	
Shanghai	67	73	4	5	71	78	
Jiangsu	1252	1428	158	203	1409	1631	
Zhejiang	704	611	177	173	881	784	
Anhui	663	435	554	507	1218	942	
Fujian	497	484	278	315	775	799	

(continued)

Table 3.3 (continued)

Province (autonomous region and municipality directly under the Central Government)	Coal equivalent (10,000 tce)					
	Commercial energy		Non-commercial energy		Total	
	2022	2014	2022	2014	2022	2014
Jiangxi	508	572	358	442	867	1014
Shandong	1200	1286	173	825	1373	2111
Henan	837	1613	37	216	875	1829
Hubei	389	558	36	271	425	829
Hunan	796	1016	32	333	828	1348
Guangdong	542	672	91	374	633	1046
Guangxi	468	350	14	177	482	527
Hainan	64	57	8	21	71	78
Sichuan-Chongqing	1431	1231	324	1021	1755	2252
Guizhou	864	742	41	221	906	963
Yunnan	203	209	112	538	315	747
Xizang	0	17	16	121	16	138
Qinghai	147	157	119	180	265	338
Ningxia	97	135	114	258	211	392
Xinjiang	963	819	23	20	986	839

(continued)

Table 3.3 (continued)

Province (autonomous region and municipality directly under the Central Government)	Coal equivalent (10,000 tce)					
	Commercial energy		Non-commercial energy		Total	
	2022	2014	2022	2014	2022	2014
Northern region	11640	13921	3011	5746	14651	19667
Southern region	8448	8438	2188	4600	10636	13038
Total	20088	22359	5215	10346	25303	32705

Note The data of Sichuan-Chongqing in the table includes Sichuan Province and Chongqing Municipality; Hong Kong, Macao, and Taiwan regions are not counted in this table due to fewer traditional rural areas of pollutants (like CO, SO_2, NO_x, and particles) and greenhouse gas CO_2. Among them, CO emissions entering rooms can easily lead to the risk of poisoning. Particles can be classified into total suspended particles (TSP) and inhalable particles by the aerodynamic equivalent diameter (referred to as particle size). TSP refers to all particles with a particle size of less than 100 μm, while inhalable particle refers to particles with a particle size of less than 10 μm, which is denoted by PM_{10}. Among them, inhalable particles with a particle size ranging from 2.5 to 10 μm are called coarse particles, and inhalable particles with a particle size of less than 2.5 μm are called fine particles, which is denoted by $PM_{2.5}$. The amount of $PM_{2.5}$ emitted from the burning of solid fuels accounts for a very high proportion of particles. $PM_{2.5}$, due to its large specific surface area, stays in the environment for a longer period, adsorbs more hazardous substances such as polycyclic aromatic hydrocarbons (PAHs) and heavy metals, and can enter the human alveoli, causing much more health hazards to the human body than coarse particles. Epidemiological research evidence shows that the smoke from the burning of coal and biomass can affect the respiratory function of the human lungs, is closely related to asthma, chronic bronchitis, chronic airway obstruction, and other respiratory diseases as well as cardiovascular and cerebrovascular diseases, and may cause multiple types of pathologic changes in the lung tissue. The thickening of endangium, the increase of atheromatous plaques, and the elevation of blood pressure are related to long-term chronic exposure to smoke from burning.

In the *2016 Annual Report on China Building Energy Efficiency*, the performance of common stoves in rural areas in burning different solid fuels is subject to laboratory and field test analyses, and their thermal efficiency and the emission factors of pollutants are obtained, as shown in Table 3.4. As can be seen from the table, the emission factors of $PM_{2.5}$ and SO_2 from the burning of bituminous coal by rustic heating furnaces are the highest, which are 3.73 g/kg dry fuel and 1.78 g/kg dry fuel respectively; in the case of burning anthracite and briquettes, the emission factors of $PM_{2.5}$ decrease successively, which are 3.33 g/kg dry fuel and 2.20 g/kg dry fuel respectively. In terms of the consumption of a unit mass of fuel, replacing the original bituminous coal with anthracite and briquette reduces $PM_{2.5}$ emissions by 10.7% and 41%, respectively. Despite the different degrees of reduction, the emission factor is still high. The emission factor of SO_2 from the burning of anthracite is the lowest, which is 0.16 g/kg dry fuel; however, as the calorific value of anthracite is the maximum, a high furnace temperature will be caused during its burning, leading to a high emission factor of NO_x. For cooking stoves, the emission factor of $PM_{2.5}$ from honeycomb briquette stoves is the lowest, which is 0.82 g/kg dry fuel; the emission factors of $PM_{2.5}$ from the burning of straw and firewood by traditional firewood-burning stoves are high, reaching 9.97 g/kg dry fuel and 8.28 g/kg dry fuel respectively, which are about 3 times of the emission factor of $PM_{2.5}$ from the burning of bulk coal; however, the emission factor of SO_2 is very small, which is only 0.02 g/kg dry fuel.

According to the emission factors in Table 3.4 and the current status of energy consumption data of rural buildings in China, it can be calculated that the total $PM_{2.5}$ emissions from domestic coal burning in rural areas of China in 2022 are 0.335 million tons and that the total $PM_{2.5}$ emissions from direct burning of biomass are

Table 3.4 Emission factors of $PM_{2.5}$ and other gas pollutants from different stoves

Stove type	Usage	Thermal efficiency (%) (Standard deviation)	Average emission factor (g/kg dry fuel)				
			$PM_{2.5}$ (Standard deviation)	CO (Standard deviation)	CO_2 (Standard deviation)	SO_2 (Standard deviation)	NO_x (Standard deviation)
10 types of rustic heating (burning bituminous coal)	Heating	32.7(9.5)	3.73(5.09)	61.05(41.17)	2497.23(103.11)	1.78(4.78)	2.05(0.37)
8 types of rustic heating (burning anthracite)		30.2(7.1)	3.33(3.29)	64.33(46.99)	2729.25(73.85)	0.16(0.19)	2.99(2.23)
8 types of rustic heating (burning briquette)		23.9(10.1)	2.20(4.43)	89.73(41.98)	2095.66(116.22)	0.30(0.35)	1.14(0.61)
3 types of honeycomb briquette stoves (burning honeycomb briquette)	Cooking	–	0.82(0.41)	53.42(49.40)	1432.23(77.62)	1.03(0.99)	0.64(0.12)
3 types of traditional firewood-burning stoves (burning straw)		–	9.97(2.01)	36.93(11.60)	1434.30(18.22)	0.02(0.02)	1.85(0.32)
3 types of traditional firewood-burning stoves (burning firewood)		–	8.28(2.46)	38.30(8.33)	1565.25(13.09)	0.02(0.03)	2.22(0.22)

0.854 million tons; the *2020 Annual Report on China Building Energy Efficiency (Special Topic on Rural Residential Buildings)* provides that, before the implementation of clean heating in 2014, the total $PM_{2.5}$ emissions from domestic coal burning in rural areas of China were 0.623 million tons and the total $PM_{2.5}$ emissions from direct burning of biomass were 1.996 million tons. It is thus clear that, by changing the energy consumption structure in rural areas from 2014 to 2022, $PM_{2.5}$, CO_2, SO_2, and NO_x emissions have been reduced by 1.43 million tons (54.6%), 130 million tons, 0.18 million tons, and 0.51 million tons respectively each year. According to the data published in the *China Statistical Yearbook on Environment 2022*, the total particle emissions, total NO_x emissions, and total SO_2 emissions in the national exhaust emissions in 2014 were 17.408 million tons, 20.78 million tons, and 19.744 million tons respectively; by contrast, the total particle emissions, total NO_x emissions and total SO_2 emissions in the national exhaust emissions in 2021 were 5.374 million tons, 9.884 million tons and 2.748 million tons respectively (Because the *China Statistical Yearbook on Environment 2023* has not been released as of the date of publication of this book, only the data of 2021 can be used for comparative analysis). Thus, it can be calculated that, during this period, the particle, NO_x and SO_2 emission reductions resulting from clean energy use in rural areas of China contribute to 11.9%, 4.7%, and 1.1% of the national total emission reduction respectively.

However, in the total $PM_{2.5}$ emission reduction in rural areas across China, the percentage contribution varies greatly from region to region. Figure 3.6 presents the percentage contribution of four regions, i.e., "Beijing-Tianjin-Hebei" and surrounding areas, other northern provinces except for "Beijing-Tianjin-Hebei" and surrounding areas, provinces in the Yangtze River Basin and other southern provinces except for those in the Yangtze River Basin, to the overall $PM_{2.5}$ emission reduction in rural areas from 2014 to 2022. It can be seen from the figure that the percentage contribution of "Beijing-Tianjin-Hebei" and surrounding areas to $PM_{2.5}$ emission reduction is the largest, reaching 49%, and the percentage contributions of other northern provinces except for "Beijing-Tianjin-Hebei" and surrounding areas, provinces in the Yangtze River Basin and other southern provinces except for those in the Yangtze River Basin are 20%, 17%, and 14% respectively. In addition, from the analysis of the CO_2 emission reduction in rural areas for the past nearly 10 years, the percentage contribution of "Beijing-Tianjin-Hebei" and surrounding areas is still the largest, reaching 82%. This fully demonstrates the remarkable effect of the clean heating campaign in northern China on energy conservation and emission reduction in this region.

3.3.2 Environmental and Health Benefits

The clean heating pilot campaign in northern China was first implemented in the "2+26" channel cities in Beijing-Tianjin-Hebei. This region is also the one with the highest proportion of clean heating transformation and with $PM_{2.5}$ emission reductions accounting for half of the total rural emission reductions in China; therefore, the

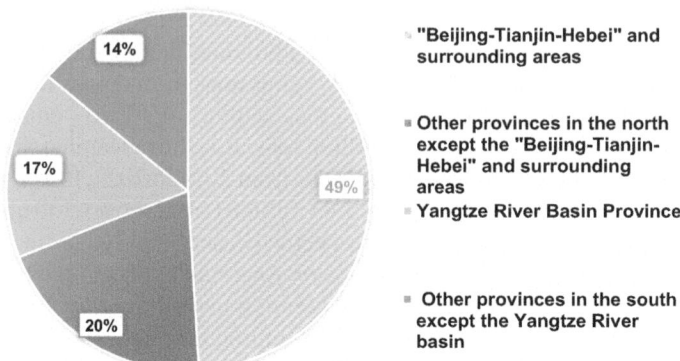

Fig. 3.6 Percentage contribution of $PM_{2.5}$ emission reduction caused by the change in domestic energy use in rural areas of different regions in China from 2014 to 2022

environmental improvement effect in this region is particularly obvious. According to the data results in the *Report on the State of the Ecology and Environment in China* released by the Ministry of Ecology and Environment of the People's Republic of China, in 2013, the ratio of the number of days meeting the air quality standard in "Beijing-Tianjin-Hebei" and surrounding areas fell within the range of 10.4–79.2%, with the average ratio of only 37.5%, and the ratio of the number of days of heavy pollution and above in the number of days exceeding the air quality standard was 20.7%, in which the number of days with $PM_{2.5}$ as the primary pollutant was the largest, accounting for 66.6%. In 2022, the ratio of the number of days with good urban ambient air quality in "Beijing-Tianjin-Hebei" and surrounding areas fell within the range of 59.2–78.4%, with the average ratio of 66.7%, up by 29.2% compared to that in 2013, and the average ratio of the number of days exceeding the air quality standard was 33.3%, in which the number of days of heavy pollution and above only accounted for 2.1%, down by 18.6% compared with that in 2013.

Figure 3.7 presents the changes in the mean annual concentration of $PM_{2.5}$ in "Beijing-Tianjin-Hebei" and surrounding areas in the 10 years from 2013 to 2022 as published in the *Report on the State of the Ecology and Environment in China*. As can be seen from the figure, the mean annual concentration of $PM_{2.5}$ in "Beijing-Tianjin-Hebei" and surrounding areas in 2013 was 106 μg/m^3. In September 2013, China launched its first action plan for the comprehensive governance of prominent environmental issues, i.e., the *Air Pollution Prevention and Control Action Plan*. Since then, by mobilizing forces of all sectors of society, China has carried out the governance of fuel coal, industrial, transportation, fugitive dust, and bulk coal pollution sources step by step on an unprecedented scale and subsequently promoted key work including the clean heating pilot work in northern China and the Three-Year Action Plan for Winning the Blue Sky War, causing the mean annual concentration of $PM_{2.5}$ in "Beijing-Tianjin-Hebei" and surrounding areas in 2022 to drop to 44 μg/m^3, down by 58.5% compared to that in 2013.

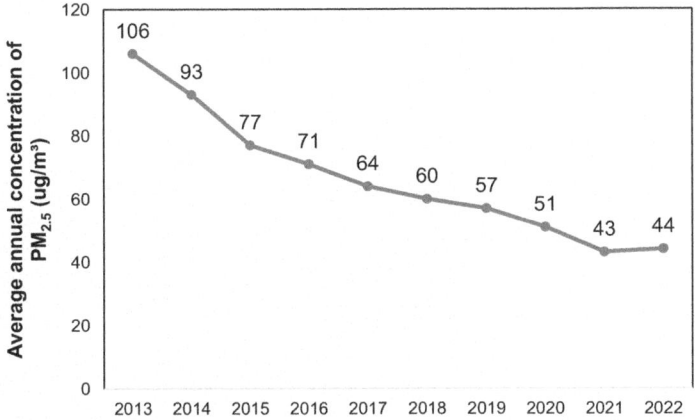

Fig. 3.7 Changes in the mean annual concentration of $PM_{2.5}$ in "Beijing-Tianjin-Hebei" and surrounding areas in the past 10 years

The large-scale promotion and application of clean energy in the northern region not only effectively improves the air quality of the regional atmospheric environment but also remarkably improves the local air quality in villages. Take a village in Fangshan District, Beijing as an example. The impact of implementing clean heating on outdoor air quality in rural areas is analyzed quantitatively by comparing the continuous monitoring data on $PM_{2.5}$ in outdoor air during the heating season in the village in 2013 and 2018. Farmer households in the village basically employed heating methods including coal stoves and heated brick beds in 2013 and changed the heating methods to low-temperature air source heat pumps in 2018. The monitoring instrument was a DASIBI 7201 type β-ray suspended particle analyzer (Fig. 3.8). The recording interval was 1 min. The monitoring was conducted on the roof of the two-floor building of the Villagers' Committee, which was relatively high in the village, thus eliminating the influence of different installation positions of instruments.

Figure 3.9 shows the changes in the daily mean concentrations of $PM_{2.5}$ in the air from January to March in the village in 2013 and 2018, and the changing trend of its test concentration was largely consistent with the average atmospheric monitoring data of Beijing. The average value of the daily mean concentrations of $PM_{2.5}$ in the air from January to March in the village was 139 μg/m³ in 2013 and decreased by 55% to 62 μg/m³ in 2018. As specified in the *Ambient Air Quality Standards* (GB 3095-2012), the Level-1 standard for $PM_{2.5}$ control concentration in the ambient air should be a daily mean concentration of 35 μg/m³ and the Level-2 standard should be a daily mean concentration of 75 μg/m³. The Level-2 standard is implemented in rural areas. By analyzing the probability density distribution of the daily mean $PM_{2.5}$ concentration in the village, it can be obtained that the daily mean concentration in 2013 was distributed at most around 120 μg/m³ while that in 2018 was distributed at most around 30 μg/m³, indicating that the daily mean outdoor $PM_{2.5}$ concentration is gradually approaching lower concentrations and that there will be more and more

Fig. 3.8 Instrument used for monitoring in a village in Fangshan District, Beijing in 2013 and 2018

days reaching the Level-2 standard and even the Level-1 standard. Implementing clean heating plays an obvious role in improving outdoor air quality.

Modernizing energy use for heating in rural areas of China can not only help achieve the dual carbon goals but also significantly improve the population's health. Research results indicate that the average life expectancy in China has increased by 2.2 years due to the improvement in air quality since 2013, to which the contribution of clean heating in rural areas reaches more than 30%. As noted in the *Clean Pathways*

Fig. 3.9 Daily mean concentrations of $PM_{2.5}$ in the air from January to March in a village in Fangshan District, Beijing in 2013 and 2018

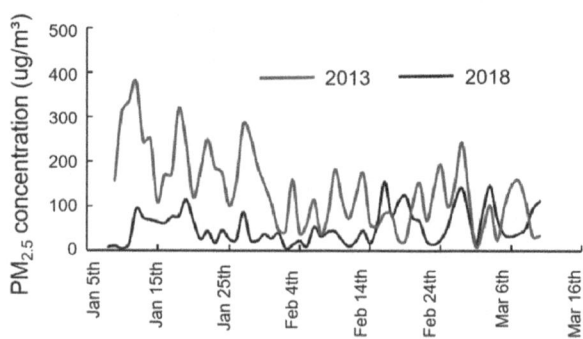

and *Environmental Health Benefits of Clean Heating and Cooking for Rural Residents in China*, a research report jointly prepared by the Institute of Carbon Neutrality, Peking University, the College of Urban and Environmental Sciences, Peking University, and the School of Environmental Science and Engineering, Southern University of Science and Technology in 2023, the improvement of domestic energy and stoves in rural areas of China from 1980 to 2014 prevented a total of 346,000 people (273,000–466,000 people) from premature death. By 2019, about 20% of rural residents (approximately 44 million farmer households) in China still used or partially used traditional solid fuels for cooking, while nearly 80% of rural residents used solid fuels for heating. Under this baseline condition, the contribution of cooking to the average outdoor $PM_{2.5}$ concentration in China was 1.1 μg/m^3, slightly higher than the contribution of heating in that year (0.91 μg/m^3). If 50% or 100% clean cooking is implemented, this contribution can be reduced to 0.66 μg/m^3 or 0.08 μg/m^3. In the future, if 50% of these farmer households switch to clean cooking, 37,000 premature deaths caused by air pollution can be prevented annually; if clean cooking and heating improvement plans are implemented simultaneously, their superimposed health benefits can prevent 69,000 premature deaths each year.

3.3.3 Analysis of Clean Heating Problems in Rural Areas

The introduction of a series of national and local policies has played an important role in the progress of clean heating. Through 5 years of implementation, from "switching from coal to electricity" and "switching from coal to gas" to the "four-appropriateness" principle (i.e., use electricity if it is the most appropriate, use gas if it is the most appropriate, use coal if it is the most appropriate, and use heat if it is the most appropriate) and from multi-energy complementation to adjusting measures to local conditions, the clean heating transformation in northern rural areas has achieved remarkable results in terms of environment, economy, especially people's livelihood. However, from the perspective of implementation, there are still some problems that affect the effect of policy implementation and the sustainable development of the industry, and targeted enhancement and improvement will be required in the future.

1. Insufficient Interdepartmental Collaboration

Clean heating transformation involves many management authorities, which do things in their own ways, one-sidedly stress their own functions, and lack the systematic consideration of policies; besides, some policies even have apparent conflicts, affecting the implementation of policies. For instance, the environmental authority mainly emphasizes the environmental protection standard and promotes clean heating but takes insufficient account of the heat source supply support managed by other authorities. In addition, to ensure compliance with environmental protection standards, some places blindly remove existing heating facilities while gas and other supplies cannot be guaranteed, affecting the residents' ability to stay warm during winter; moreover, the boundaries of responsibilities are not clear among local governmental departments, causing the implementation of grass-roots work to be not very smooth.

2. Undifferentiated Treatment of Fiscal Subsidies

The central fiscal subsidy policies fail to fully reflect the difference in the transformation workloads of pilot cities, and the difficulty in the implementation varies greatly among pilot cities. The pilot cities have different transformation workloads, and the gap is large among some cities. Currently, central subsidies are paid in fixed amounts according to administrative levels. Pilot cities at the same administrative level still receive the same amount of subsidies although there may be a large gap in their workloads. For example, among the same batch of pilot cities, a prefectural-level city with the maximum transformation workload plans to transform the heating methods of 1.3271 million households in three years, and a city with the minimum workload plans to transform the heating methods of 0.1616 million households in three years. The workload of the former is 8.2 times that of the latter, but the two cities receive the same amount of central subsidies. This leads to a great difference in the central fiscal subsidies that the households in the pilot cities can enjoy.

Furthermore, the subsidy policies fail to take full account of the differences in the costs of using different technologies. In most pilot cities, the different types of "coal to electric heating switching" and "coal to heat pump switching" technologies enjoy the same subsidy standard, which is even consistent with the maximum subsidy amount for "switching from coal to gas".

3. Unclear Renewable Energy Development Policies

The economic incentive policies for heating with renewable energy are not clear enough, and the subsidies are insufficient. In the pilot cities and projects for clean heating in northern China, most cities have formulated clear subsidy policies for "switching from coal to electric heating" and "switching from coal to gas." The subsidies cover one-time investment (equipment subsidies) and operation subsidies, in which the equipment subsidies for "switching from coal to electric heating" and "switching from coal to gas" generally exceed 50%. Only a few cities have formulated targeted incentive policies to promote the application of biomass, solar energy, terrestrial heat, air energy, and other renewable energy in the heating field, and most of them only refer to the subsidy policies for "switching from coal to gas" or "switching from coal to electric heating" and lack clear targeted subsidy policies and standards.

4. Lack of Scientific Guidance for Planning and Appropriate Technical Routes

In terms of existing policies, most of them only issue task objectives and list technologies and lack in-depth and systematic feasibility analysis and comparison of the technologies based on local reality and rigorous analysis and demonstration of the selected technical routes. As a result, some pilot cities have an unclear understanding and knowledge of clean heating; in the implementation process, such cities focus the work on completing projects and tasks and lack detailed guidance and management for each link, which may cause differentiated results like fuzzy performance and contrasting energy-saving effects of the different technical solutions. In addition, many pilot cities emphasize the construction link and lack comprehensive consideration of the subsequent operation and maintenance.

5. Vague Implementation Scheme for Clean Heating Transformation

During the advancement of the clean heating transformation project, because of a tight schedule and a heavy task, relevant departments have formulated implementation schemes that are too general, making clean heating transformation schemes and technical routes extremely similar. The simple "switching from coal to electric heating" and "switching from coal to gas" are generally predominant in rural areas, and no targeted design has been made by fully combining the rich biomass resources including solar energy, crop straw, branches from fruit tree pruning and livestock and poultry manure as well as infrastructure and affordability of the common people in the vast rural areas in northern China. Besides, many pilot cities have neglected building energy efficiency retrofitting in rural areas. The building envelopes of existing buildings in rural areas generally have poor thermal insulation properties. If energy efficiency retrofitting is not carried out accordingly, the heating effect will be poor and the energy consumption will be high, thus directly affecting the enthusiasm of farmer households to participate in clean heating.

6. Imperfect Market-Oriented Operation Mechanism

The market-oriented operation mechanism of clean heating is imperfect, and currently, the main form is government investment, which has the problem of difficult and expensive financing. Clean heating is primarily driven by the government and depends on direct government input. Some areas introduce social capital (such as thermal, power, and gas enterprises) in the franchise or Public–Private Partnership (PPP) model to invest in construction and operation and maintenance. Due to the low profitability of clean heating projects and low market enthusiasm, the operation model in which enterprises assume the main responsibility and governments provide support to ensure affordable heating for the people has not been truly established.

At present, although there are abundant financial support tools such as green credit, credit funds, policy-based loans, and finance leases in the market, the main force of financial support for clean heating is still bank credit and the form of support is single. Restricted by collateral, clean heating projects have the problem of difficult and expensive financing.

7. No Established Sustainable Development and Long-Term Market Mechanism for Clean Heating

Currently, the switching from coal to clean energy relies on the strong executive power of the government and is promoted from top to bottom. The national and local fiscal subsidies at all levels cover most of the cost of project construction and are still being continuously provided for the operation and use of residents. The market-oriented, government-driven, and affordable-for-residents promotion model is still not formed, and the long-term market mechanism for clean heating is to be established.

The cost of using clean heating is generally rising. After the cancellation of subsidies, heating expenditures are less affordable for rural residents. Compared with heating with bulk coal, clean heating in rural areas has higher usage costs; even after rural residents enjoy price subsidies, their heating expenditures are still generally rising.

8. Assessment Mechanism Emphasizing "Quantity" Over "Quality"

Currently, "quantity" is taken as the main assessment indicator for clean heating, and the influence of "quality" is neglected. Most areas have not established a data operation monitoring platform for clean heating projects in winter and lack quantitative data for objective evaluation of the implementation, making it difficult to timely correct the technical routes. At this point, supervision and assessment focus on clear observation and lack comprehensiveness and objectivity to a certain extent, thus making it easy to mislead the further improvement and optimization of relevant policies.

3.4 Analysis of Heating Modes and Energy-Saving Paths of Northern RRBs

Significantly different from the concentrated population in cities, the rural population is scattered, and a small-scale settlement mode has been formed with villages as the main administrative units. For a long time, most rural residential buildings have been self-built and are mainly scattered single residential buildings lacking overall planning and construction standards. Residents in a village learn from each other in terms of residential building form and energy consumption, which even tend to be similar. Therefore, building energy efficiency work in rural areas also needs to be carried out with villages as the basic units.

In addition, there are significant differences between rural and urban residential buildings in many aspects such as historical traditions, land resources, production methods, lifestyles, and natural conditions. The building form, population composition, inherent living patterns, types of people's activities, resource characteristics, and economic levels in rural areas determine the difference in building use patterns, behavioral patterns, and indoor thermal environment needs of rural population and concentrated urban population.

To do a good job of clean heating in northern China, we first need to understand the actual heating demand and characteristics of rural residential buildings. The following describes the main characteristics of RRB in northern China, the heating-related behavioral characteristics of users, and the energy-saving paths suitable for these characteristics from three aspects.

3.4.1 "Partial-Time, Partial-Space" Mode is a Common Characteristic of RRB Heating

1. The Rural Production Mode Determines the Scattered Settlement Pattern and Large Floor Area of RRBs

For most rural areas in China, agriculture (forestry and animal husbandry) and other decentralized activities still dominate farmers' production. To ensure the sufficient utilization of arable land resources, rural residents tend to live dispersedly and independently in villages, manage land in different areas, and supplement it with cottage and breeding to make their living. This mode of production determines the unique residential pattern of farmer households and the usage of RRBs: Rural residential buildings are not only the living space of farmers but also their important production and auxiliary space. For example, farmer households must have enough indoor space for the storage of grain produced by themselves; moreover, they need to have enough space in the courtyard to store farm implements, tractors, and other production equipment; they also need to carry out small-scale production activities such as vegetable cultivation, poultry breeding, handicraft production and basket weaving either in the courtyard or indoors. In addition, the multi-generational residential pattern is commonplace in rural residential buildings, further increasing the diversity of farmer households' demand for spatial functions and service levels within RRBs. Therefore, RRBs need to meet the multiple requirements of different activities and populations, and one of the important characteristics of rural residential buildings is the combination and unification of production and living functions.

On the contrary, after years of development, the production space and living space have been completely separated in urban areas, and the functional design of urban residential buildings only needs to satisfy the living demand, without considering the production demand.

The production and the functional characteristics of rural residential buildings dictate that rural areas must maintain the scattered settlement pattern and cannot adopt the centralized pattern of urban areas. Ensuring that the homesteads of RRBs have a large floor area and are equipped with an independent courtyard is a prerequisite for guaranteeing daily life and economic production activities of farmers, which also determines the particularity of rural areas in terms of land use and architectural space planning for residential buildings. In this mode of life and production characterized by scattered living and self-sufficient land management, the per capita land occupancy is generally much higher than the level of urban areas, despite an imbalance in the regional distribution, and the population density is relatively small; moreover, the rural residential buildings generally are single-floor or low-rise buildings and independent courtyards. The peripheral land resources and internal space of buildings in rural areas are relatively abundant, and the overall land use for construction of RRBs and their internal layout are relatively loose.

As can be seen from the common urban residential building floor plan and rural residential building floor plan in Fig. 3.10, a unit in urban high-rise and medium–high-rise concentrated residential buildings is relatively compact, and the access between

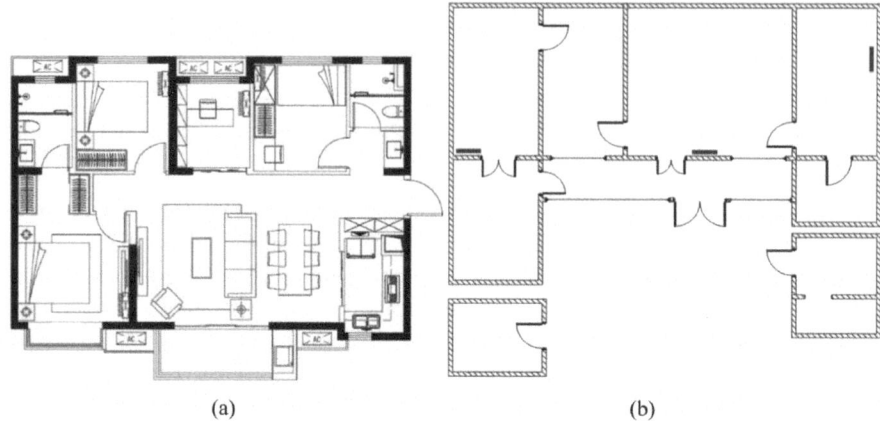

Fig. 3.10 Comparison of floor plans of common urban and rural residential buildings (m). **a** Urban residential building; **b** Rural residential building

the interior of the unit and the outdoor area is usually only few exterior windows and an exterior door. Due to the large number of floors, the shape factor is small and the building energy consumption is low. In contrast, RRBs are mostly a single-floor independent building with a courtyard. The main rooms face south, and rooms of different functions often have an open-air courtyard as the intersection and link of activity trajectories. To reach different functional rooms, it is often necessary to pass through the outdoor space. Households with better conditions will add a sunroom in the south direction, thus utilizing solar radiation to raise the indoor temperature and improve the airtightness of the building at the same time. As there has been no relevant construction standard for the overall planning of rural residential buildings for a long time, most of them are built by farmers based on local climatic characteristics and technical conditions as well as historical traditions, and the thermal performance of the building envelope is generally poor.

2. Lifestyle and Economic Level Determine the Special Behavioral Pattern and Heating Demand of Farmer Households

The scattered settlement pattern in rural areas is the foundation of daily life and economic production activities of farmers, and a relatively abundant homestead is the necessary guarantee for their production and life. Moreover, the unique construction form of RRBs and the traditional lifestyle of farmers make their requirements for the indoor environment comfort level and service level of RRBs quite different from those of urban residents.

In northern China, the current design temperature of winter heating for urban residential buildings is 18 °C, but most residents expect a comfortable room temperature of 20 °C or even higher. This temperature requirement is consistent with the fact that urban residents go in and out of their houses less frequently every day and do not need to change their clothes when entering and exiting different rooms. However, in rural areas, due to production and living habits, people frequently go in and out of their houses and usually spend most of their time in the living room when they are in their houses. As indicated by the results of the research on the daily activity pattern of a typical farmer household in the suburbs of Beijing, the members of this household stay in the living room for 70% of the time during the 11 h of the day (7:00–18:00), but they enter and exit the living room 16 times every day in the daytime and the duration of staying outside the living room varies from 2 to 60 min each time. Such frequent entry and exit will cause great inconvenience to the life of the farmer household if they change their clothes every time they enter or leave the room. Therefore, the clothing level of farmer households follows the criterion that they will not feel cold in short-term outdoor activities, which determines that the design temperature of winter heating for RRBs is lower than the urban heating temperature. The results of a large number of studies also show that, in the opinion of most farmers in northern China, the difference between indoor and outdoor temperature should not be too large in winter and the average temperature of the living room and bedroom should be 5–6 °C lower than that in cities.

The traditional heating methods in rural areas that have evolved along with the abovementioned living habits and heating demands are significantly different from the heating methods of urban buildings. Correspondingly, the charging method for heating in rural residential buildings is also quite different from that for central heating in urban residential buildings. The urban multi-story and high-rise residential buildings have highly concentrated habitable space, the buildings have a small shape factor, the heat supply network has high thermal inertia, and most of the urban residences using central heating systems maintain an indoor temperature ≥ 18 °C and use the heat supply model that maintains continuous heating for interior space and charges heating fee by the heating area (some regions are trying the model that charges heating fee by the floor area and actual heat supply). The rural residential buildings are stand-alone buildings, the heating system of the RRBs is controlled by the farmer households, and they will decide whether to and when to use heat supply according to their living habit, awareness of energy conservation, and cost of heat supply. This model allows rural households to conserve energy by autonomously adjusting their behaviors.

There is a distinct difference between rural residents and urban residents in terms of their requirements for room temperature. As such, the indoor thermal environment control objectives of rural residential buildings must be established based on the individual needs of the rural dwellers, the interior temperature is allowed to change between day and night, and not every room needs heating. Due to the temperature difference between day and night, the thermal insulation of building envelope, window-to-wall ratio, and other building energy efficiency parameters of rural buildings will differ greatly from those of the urban buildings, and thus must be carefully analyzed according to the reality of the rural areas. Overall, the lower interior temperature and shorter heating duration in winter are advantages of RRBs in improving building energy efficiency, and this model should be advocated and encouraged.

3. Rooms of Different Functions in RRBs Require Different Levels of Heating

Due to the architectural characteristics and living habits of rural households, the users stay for different durations and perform different activities in the rooms of different functions in RRBs, and thus the rooms require different levels of heating. Figure 3.11 illustrates a test conducted by Tsinghua University regarding the heating needs of different rooms of a typical rural household in Beijing. This household has a 223 m^2 floor area, a 158 m^2 heating area, and 5 family members. The permanent residents are two adults and one child, while the other two members only come home on holidays and some weekends. The rooms of this household have clearly differentiated functions, which dictate their heating needs. The rooms including the living room, master bedroom, kitchen, dining room, bathroom, middle bedroom, guest bedroom, and west bedroom are all equipped with flexible and separate air-to-air heat pumps with low ambient temperature. The operation of such a heating system is not supported by any subsidies, and the operation model is completely determined by the needs of the user. As indicated by measurement of the temperature and humidity in the rooms, the supply and return air temperature of the air-to-air heat pump, and power consumption of the air-to-air heat pump, the rooms differ markedly in terms of heating duration in the heating season (expressed as heating utilization rate, i.e., the ratio between the daily average time of using the heating equipment in winter and the total time of the day) (Fig. 3.11c, d). The living room has the highest heating utilization rate, which is 50.8%. There are plants grown in the living room, so the air-to-air heat pump has to keep working even at night, and the heating duration is 20:00–8:00. As the temperature rises in the daytime, the user turns off the air-to-air heat pump. The heating utilization rate of the commonly used bedroom is 15.2%. During sleeping time, the user prefers to keep a temperature not too high, and usually turns on the air-to-air heat pump the half hour before sleep and upon wake-up. The average temperature in the living room at night is <18 °C. The kitchen, dining room, and bathroom have heating utilization rates of only 13%, 8.6%, and 7.1%, while the other rooms have less than 6%.

Besides, the rural areas are losing the labor force. Those who live in the RRBs are mostly old people, while the young and middle-aged are studying or working in other places and only return home on holidays. Each rural household has many rooms, but most of the rooms are vacant and do not need a heat supply. The rooms that are occasionally inhabited only need temporary heating. This is the reality of the rural areas in China and explains why they use a special heat supply method.

As such, the rural areas differ greatly from the urban areas in terms of mode of production, land resources, residence usage mode, frequency of usage of different rooms, and indoor thermal environment. The intrinsic characteristics of rural areas dictate the special heat supply model of the rural residential buildings. Therefore, the building energy efficiency strategy and energy conservation technology for the rural areas should not copy those of the rural areas, and the heat supply model needs to be rationally guided and designed to achieve building energy efficiency of RRBs and improve the indoor thermal environment, so as to establish a sustainable development model that serves the actual needs of the rural areas. Overall, low heating demand in

3.4 Analysis of Heating Modes and Energy-Saving Paths of Northern RRBs

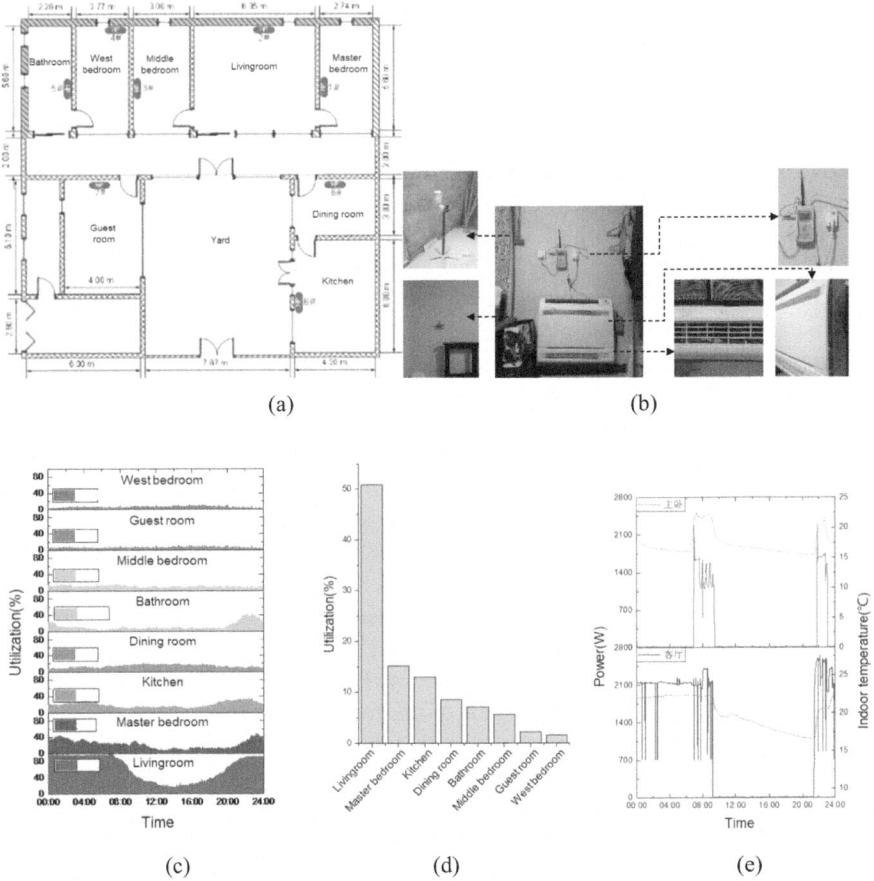

Fig. 3.11 Measured actual heating needs of different rooms of a rural household in Beijing. **a** Typical floor plan; **b** onsite test diagram; **c** time-dependent operation rates of the air-to-air heat pumps in the rooms; **d** heating utilization rates of the rooms; **e** power of the air-to-air heat pumps in the main heated rooms on typical days and the interior temperature curve

winter and diversified usage of different architectural spaces are not the objectives we consciously pursue, but the beneficial factors that help the RRBs conserve energy.

3.4.2 Rational Thermal Insulation of the Building Envelope of RRBs is the Fundament of Building Energy Efficiency

Most RRBs in North China are built by the rural dwellers themselves, so the design and construction quality of these buildings is usually quite low and the rooms are uncomfortable to inhabit. The building materials, structural form, and thermal performance of the external envelope parts of RRB such as the roof, exterior walls, and exterior windows will significantly affect the heating energy consumption and indoor thermal environment in winter. Currently, most rural buildings in North China do not have thermal insulation, the exterior walls are too thin, and the exterior doors and windows have poor thermal performance and air tightness. The RRBs also have a large shape factor. Therefore, the heating load of the RRBs in winter is 2–3 times that of urban buildings. Now the heating of RRBs in winter does not have such high-energy consumption, mainly because the interior temperature is low, and most RRBs do not receive heating the same way all the rooms of urban buildings receive continuous heating in winter.

Rational thermal insulation of building envelope will reduce the heating energy consumption and heating fee, and also significantly improve the interior thermal comfort, so it is the fundamental of rural building energy efficiency.

The current energy efficiency retrofitting of rural houses is mostly overall thermal insulation, i.e., external thermal insulation of four exterior walls, roof thermal insulation, and installation of plastic steel double-glass window or aluminum alloy windows. Take a northern city, for example. The assessment of the energy efficiency retrofitting of over 1 million rural houses indicates that, after the retrofitting, the average indoor temperature in winter increases by 4–7 °C, and the heating coal consumption reduces to 11.6–15.1 kgec/m^2, which is 27–44% less than it was before the retrofitting. This is great energy efficiency. However, this retrofitting does not take account of the actual heating needs of the rural households, e.g., thermal insulation materials are installed in the rooms that do not need heating. In addition, the thickness and varieties of the thermal insulation materials are not considered, and a one-size-fits-all approach is used in this regard. Most importantly, the cost performance of such retrofitting is too low, for the average retrofitting cost of the households is RMB 15,000–20,000. This approach cannot be widely applied in regions that are not wealthy, which is the main reason building energy efficiency retrofitting is hard to implement in rural areas.

According to a field investigation, most RRBs in the northern rural areas only have 1–2 rooms that need heating. Compared with the overall thermal insulation solution, the targeted thermal insulation of the building envelope of the commonly used rooms will realize precision energy conservation and substantially reduce the retrofitting cost, thus making the retrofitting easier to implement. Besides, the variety and thickness of thermal insulation material will determine the effect of energy efficiency retrofitting. Usually, lower thermal conductivity and greater thickness of the thermal insulation material bring better thermal insulation results, but come with a higher cost. Therefore, there is a need to strike a balance between the thermal insulation effect and retrofitting cost of RRBs. From only the perspective of energy efficiency, the retrofitting should use thick materials with excellent thermal insulation properties. With consideration of the financial capacity of the rural households, however, the retrofitting cost should not be increased excessively. For the RRBs in question, rational analysis and planning are needed to determine the thermal insulation method

3.4 Analysis of Heating Modes and Energy-Saving Paths of Northern RRBs

of the building envelope and to maintain the balance of thermal insulation effect and retrofitting cost.

Given that, the menu-based targeted thermal insulation retrofitting solution is proposed and is explained below.

1. Menu-Based Targeted Optimization

The menu-based targeted optimization method (hereinafter referred to as "menu-based method") means the RRB retrofitting options are offered like the options on a menu. The users may select the optimum retrofitting option according to their budget and needs. The menu-based retrofitting solution effectively maintains a good balance between option selection, energy conservation effect, and cost through the linear programming algorithm.

Figure 3.12 shows the flow diagram for this method. First, the energy consumption is created by using the investigation information, and the basic energy consumption of the RRBs is determined through simulation. Then the energy consumption of the retrofitting solutions (with different materials and thickness) for the exterior door, roof, exterior doors and windows, and interior partition (partition between heated room and unheated room) is dynamically simulated, to obtain the separate "menu options". Then the "menu options" are combined to create different "packages." Now the "menu" database that contains the best solution, worst solution, and general solution is created, and each package has a certain cost and energy conservation rate. The last step is optimization through the linear programming algorithm. The budget cost is input, to obtain the best solution under the cost, i.e., the "recommended" package. The user may select the "options" for the roof and exterior wall or the combination thereof. After selection, the system will present the quantified cost and energy conservation effect of the selected package.

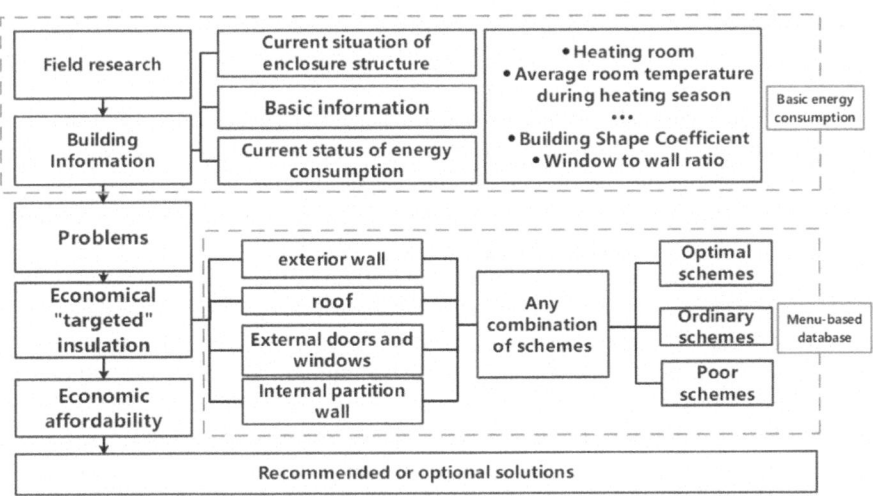

Fig. 3.12 Flow diagram for menu-based energy efficiency retrofitting of building envelope

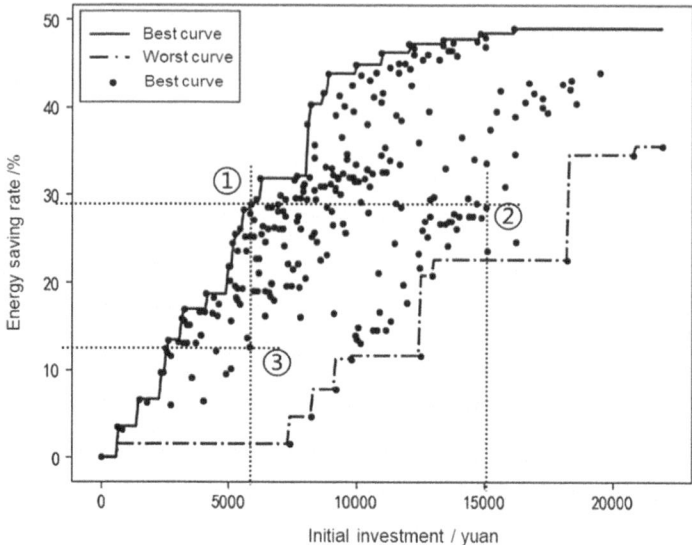

Fig. 3.13 Comparison of the results of menu-based energy efficiency retrofitting solutions

The results of menu-based energy efficiency retrofitting solutions are shown in Fig. 3.13. Each point in the figure represents a "package," i.e., a retrofitting solution. The abscissa represents the initial investment of the solution, and the ordinate the energy conservation rate. The figure contains the upper and lower envelope lines. The point on the upper envelope line represents the best solution under the given cost, while the point on the lower envelope line represents the worst solution. The two envelope lines rise sharply and then become gentle. This indicates that the investment and benefit of the thermal insulation are in non-linear correlation. If the lines pass a critical point, the increased investment will not bring a noticeable energy conservation effect.

Besides, the figure also indicates the differences between the solutions. For example, solution ① is 30 mm thick phenolic aldehyde internal thermal insulation + insulated curtain on east, west, and north exterior walls, and solution ② is 30 mm thick polyurethane internal thermal insulation + broken bridge aluminum double-glass window on east, west, and north exterior walls.

Solution ③ is external thermal insulation of 80 mm expanded polystyrene board on the north wall + aluminum alloy double-glass window. Solutions ① and ② can realize a 30% energy conservation rate, but their investments differ by nearly 3 times. Similarly, solutions ① and ③ have the same initial investment, but their energy conservation effects differ by 15%. Among the aforementioned three solutions, solution ① is clearly superior to the other two solutions in terms of retrofitting cost and energy conservation effect. As can be seen, the menu-based targeted optimization method will help the decision maker quickly identify the best solution and thus avoid ineffective investment.

3.4 Analysis of Heating Modes and Energy-Saving Paths of Northern RRBs

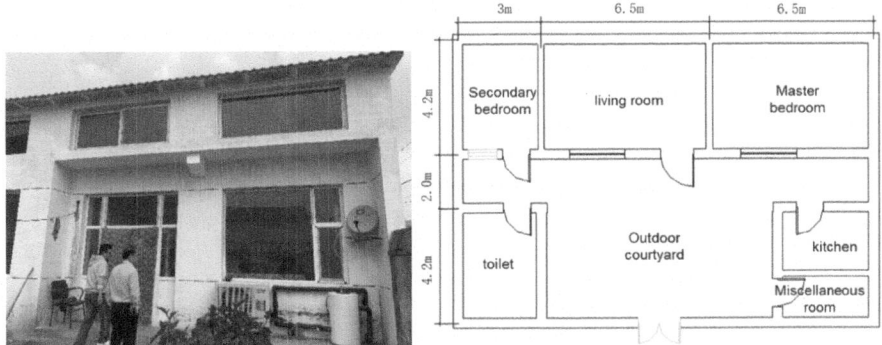

Fig. 3.14 Plan of an RRB in Shanxi

Now, two typical RRB cases in North China and Northeast China are used as examples to explain how to use this method.

2. Cold Area in North China—A Typical RRB Case in Shanxi

This RRB (Fig. 3.14) is a one-and-half-story masonry-timber structure. The ground story is living space, and the half story above is storage space.

The exterior wall is a 370 mm uninsulated solid brick wall, and the exterior window is a single-glass plastic steel window. The heating method of the household is a coal-fired furnace + heated kang, and the heated rooms are the master bedroom and living room, with a heating area of 55 m^2. The coal consumption in one heating season is approximately 3 t and the heating fee is RMB 4,000–5,000. The local rural households use continuous heating and intermittent heating, so simulation and calculation are done for these two scenarios. In the intermittent heating scenario, the heating duration of the living room is 8:00–20:00, and that of the bedroom is 20:00–22:00. The setting and calculation results are shown in Table 3.5.

Based on the concept of "targeted thermal insulation," the thermal insulation options available to this RRB include internal thermal insulation on the north exterior wall of a heated room, overall external thermal insulation on the north exterior wall, internal or external thermal insulation on east and west exterior walls, ceiling thermal insulation of ground story, and installation of the insulated curtain. Internal thermal insulation on walls will not occupy much interior space, and the thermal insulation thickness is limited to 50 mm. Through menu-based targeted optimization calculation, the 5 retrofitting solutions with 10–50% energy conservation rate and high economic efficiency in the continuous heating and intermittent heating scenarios are determined, as shown in Table 3.6.

As seen in Table 3.7, with the same thermal insulation solution, the continuous heating mode achieves a higher energy conservation rate and heat consumption than the intermittent heating mode. Whichever heating mode is used, the internal thermal insulation on the exterior wall has better cost performance than external thermal insulation. Take the north exterior wall, for example. Since only two rooms need heating,

Table 3.5 Information on an RRB in Shanxi

Building information		
Heated room	Master bedroom and living room	55 m²
Heating time	November 1 next March 31	
Heating method	/ Some of the spaces use continuous heating/intermittent heating	
Interior design temperature in the heating season		18 °C
	Building envelope	
Exterior wall	370 mm solid brick wall	1.5 W/(m²·K)
Interior wall	240 mm solid brick wall	2.0 W/(m²·K)
Exterior window	Single-glass plastic steel window	5.7 W/(m²·K)
The roof of a small second floor house	Cement tile + load-bearing wood purlin + 50 mm straw board	1.0 W/(m²·K)
The roof of first floor house	Reinforced concrete floor slab	3.1 W/(m²·K)
Building energy consumption–continuous heating		
Energy consumption index in the heating season	198.2 kWh/m²	
Heat consumption index in the heating season	68.8 W/m²	
Building energy consumption–intermittent heating		
Energy consumption index in the heating season	107.3 kWh/m²	
Heat consumption index in the heating season	63.9 W/m²	

only the north exterior walls of these two rooms need thermal insulation. However, external thermal insulation on walls needs to preserve the pleasant appearance of the RRB, so the north wall of the west bedroom, which is uninsulated also needs to be modified, which will increase the retrofitting cost. Internal thermal insulation will avoid such ineffective investment.

3. Severe Cold Area in Northeast China—A Typical RRB in Heilongjiang

The plan of the RRB is shown in Fig. 3.15. This RRB is a three-and-half-room structure comprised of one living room, two bedrooms, and one kitchen. Its exterior wall is a 500 mm thick solid brick wall, and behind it is a storage space with a 370 mm thick solid brick wall. The exterior window is a plastic steel double-glass window. The roof has load-bearing wood purlins, the rooftop is covered with reed mat and cement, and the outermost layer is a prepainted steel plate. In winter, the two bedrooms and the living room of the RRB need heating, which is a coal-fired furnace + heated kang, and the fuels are bulk coal and corncob. The RRB consumes approximately 1.5 t of bulk coal and about 7 t of corncob in the heating season. The detailed information is given in Table 3.8.

According to the investigation, the local residents obtain heat supply by building a makeshift sunroom in winter, as shown in Fig. 3.16. Most sunrooms are built to the south of the building and resemble a plastic greenhouse. Its supports are steel pipes,

3.4 Analysis of Heating Modes and Energy-Saving Paths of Northern RRBs

Table 3.6 Energy efficiency retrofitting solutions for an RRB in Shanxi (continuous heating)

	Insulation measures	Energy conservation rate (%)	Initial investment (RMB)	Energy consumption index (kWh/m^2)
Original/contrast	Uninsulated	0	0	198.2
Solution 1	Thermal insulation of 30 mm thick polyurethane board on east exterior wall + insulated curtain	20	1476	158.6
Solution 2	External thermal insulation of 50 mm thick EPS on east and west exterior walls + insulated curtain	24	2529	150.6
Solution 3	Internal thermal insulation of 40 mm thick phenolic aldehyde on the north and east exterior walls	35	3478	128.8
Solution 4	Internal thermal insulation of 40 mm thick phenolic aldehyde on the north exterior wall + internal thermal insulation of 30 mm thick phenolic aldehyde on the east exterior wall + insulated curtain	37	4000	124.9
Solution 5	Internal thermal insulation of 30 mm thick phenolic aldehyde on the north exterior wall + external thermal insulation of 50 mm thick EPS on the east exterior wall + insulated curtain	39	4978	120.9
Solution 6	Internal thermal insulation of 50 mm thick polyurethane on the north exterior wall + external thermal insulation of 60 mm thick EPS on east and west exterior walls + insulated curtain	50	6547	99.1

Note The original RRB has a 370 mm solid brick wall, single-glass plastic steel exterior window, and uninsulated pitched roof, and is a small two-story building. The material prices are known from market research and the construction cost from the local projects survey. The service life of polyurethane board, phenolic resin board, and expanded polystyrene board (EPS) is 20 years, and that of insulated curtain is 5 years

Table 3.7 Energy efficiency retrofitting solutions for an RRB in Shanxi (intermittent heating)

	Insulation measures	Energy conservation rate (%)	Initial investment (RMB)	Energy consumption index (kWh/m^2)
Original/contrast	Uninsulated	0	0	107.3
Solution 1	Internal thermal insulation of 30 mm thick phenolic aldehyde on the north exterior wall	13	2449	93.4
Solution 2	Internal thermal insulation of 30 mm thick phenolic aldehyde on the north exterior wall + insulated curtain	16	2998	90.1
Solution 3	Internal thermal insulation of 40 mm thick phenolic aldehyde on the north exterior wall + internal thermal insulation of 30 mm thick phenolic aldehyde on the east exterior wall + insulated curtain	20	3999	85.8
Solution 4	Internal thermal insulation of 40 mm thick polyurethane on the north exterior wall + internal thermal insulation of 30 mm thick phenolic aldehyde on the east exterior wall + insulated curtain	22	5368	83.7
Solution 5	Internal thermal insulation of 40 mm thick phenolic aldehyde on the north exterior wall + thermal insulation of 30 mm thick phenolic aldehyde on ceiling + insulated curtain	28	5842	77.3

Note The original RRB has a 370 mm solid brick wall, single-glass plastic steel exterior window, and uninsulated pitched roof, and is a small two-story building. The material prices are known from market research and the construction cost from the local projects survey

and the top is a plastic film. The sunrooms increase the air tightness of the RRB, and improve the heating effect through the greenhouse effect.

Figure 3.17 shows the makeshift sunroom temperature and the outdoor temperature of a rural household measured on March 23, 2023. As can be seen, the temperature in the sunroom in the daytime is 30 °C maximally and 0 °C minimally, and is 3–26 °C higher than the outdoor temperature. According to the simulation and calculation, the heating by the sunroom will decrease the energy consumption of the RRB by 10–16%.

3.4 Analysis of Heating Modes and Energy-Saving Paths of Northern RRBs

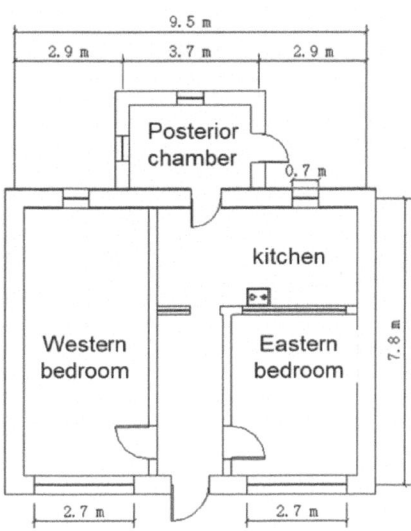

Fig. 3.15 Plan of an RRB in Heilongjiang

Table 3.8 Information on an RRB in Heilongjiang

Building information		
Heated room	West bedroom, east bedroom, living room	66 m²
Heating time	October 15 next April 15	
Heating method	Some of the spaces use continuous heating	
Interior design temperature in the heating season		14 °C
	Building envelope	
Exterior wall	500 mm solid brick wall	1.2 W/(m²·K)
Interior wall	240 mm solid brick wall	2.0 W/(m²·K)
Exterior window	Plastic steel double-glass window	2.2 W/(m²·K)
Roof	Prepainted steel plate + mortar + reed mat + wood purlin + calcium-plastic board ceiling	0.9 W/(m²·K)
Building energy consumption		
Energy consumption index in the heating season		174.8 kWh/m²
Heat consumption index in the heating season		39.3 W/m²

Fig. 3.16 Makeshift sunroom of an RRB in Northeast China

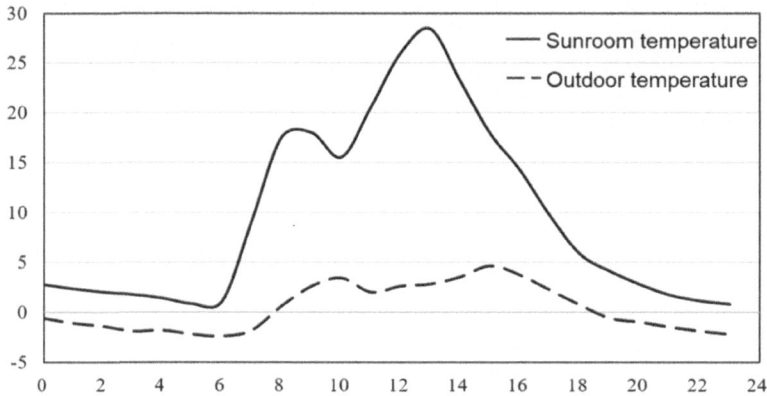

Fig. 3.17 Comparison between makeshift sunroom temperature and outdoor temperature

The makeshift sunroom solution is included in the "exterior doors and windows" sub-solutions library in the menu-based energy efficiency retrofitting database. After optimization calculation, the 5 comprehensive retrofitting solutions with 10–50% energy conservation rate and high economic efficiency are obtained, as listed in Table 3.9.

As can be seen from the above results, the makeshift sunroom has better economy and energy saving, playing a significant role in energy saving. Coupled with the external thermal insulation of extruded polystyrene board on the east, west, and north exterior walls, a 38% energy-saving effect can be achieved. To control the energy consumption index within 100 kWh/m^2, about RMB 7,000 is additionally required to do thermal insulation for the roof and external walls of heating rooms.

3.4 Analysis of Heating Modes and Energy-Saving Paths of Northern RRBs

Table 3.9 Energy efficiency retrofitting solution for an RRB in Heilongjiang

	Insulation measures	Energy conservation rate (%)	Initial investment (RMB)	Energy consumption index (kWh/m^2)
	Uninsulated	0	0	174.8
Solution 1	Makeshift sunroom	16	1200	146.8
Solution 2	40 mm thick XPS external thermal insulation on the north exterior wall + makeshift sunroom	20	2529	139.8
Solution 3	Makeshift sunroom + 30 mm thick phenolic aldehyde board thermal insulation on the roof	31	4296	120.6
Solution 4	40 mm thick XPS external thermal insulation on east, west, and north exterior wall + makeshift sunroom	38	5970	108.4
Solution 5	40 mm polyurethane board insulation in the northern wall of the west bedroom and the eastern wall of the east bedroom + makeshift sunroom + 40 mm roof insulation with phenolic foam board	44	7281	97.9

Note The price of relevant materials is derived from market research, and the construction cost from the local actual project research; the extruded polystyrene board (XPS), phenol resin laminate, and polyurethane panel feature a life of 20 years, while the plastic film for the makeshift sunroom is replaced once every two years

3.4.3 Flexible Regulation of Heating Equipment is the Core of Building Energy Efficiency

Under the actual demand mode of "partial-time, partial-space," it is an important energy-saving way to reduce energy oversupply based on securing user requirements. There is an interplay between the actual heating demand characteristics and heating equipment and system. For example, the radiant floor heating terminal system of continuous heating is just suitable for users with continuous heating requirements; for users who go out early and return late, are not at home at noon, and do not need continuous heating, the heat supply in the noon hours is ineffective heat supply. However, the system does not stop immediately. This part of the ineffective heat supply cannot be avoided, and it will also cause the illusion that users require continuous heating. Without enough knowledge about the interaction between actual demand and supply, inappropriate heating methods would be chosen. The difference between supply and demand can lead either to excessive heating and energy waste, or to the fact that the indoor thermal environment does not meet the user requirements. At present, the level of residential energy consumption in China is far lower than that of developed countries such as the United States, which is closely related to the frugal on-demand usage pattern formed by Chinese residents for a long time. With the large-scale efforts to advance clean heating in northern rural areas, it is of great significance to choose the appropriate heating technology according to local conditions. This helps prevent building energy consumption from the mode of high-energy consumption.

Subject to poor adjustability, urban central heating, radiant floor heating, and other methods cannot be adjusted for the heating characteristics of "partial-time, partial-space" in rural residential buildings, and should not be simply copied. Even with an efficient heat source like household air source or ground source heat pumps, the heating costs of users are not really reduced without overall consideration of the actual heating demand and dynamic characteristics of the RRBs. Seen from the building form, occupant type, and income level, as well as the actual building space usage pattern, thermal comfort characteristics, and heating demand of RRBs, the "urbanization" mode of "full-time, full-space" is not applicable to rural residential buildings in northern China. In contrast, the heating mode of "partial-time, partial-space," featuring room-specific installation and flexible regulation of heating equipment, can better match the actual load characteristics of RRBs, and achieve the clean heating of the vast rural areas in northern China with the lowest energy consumption.

Specifically, the heating system and heating equipment should meet the following requirements to enable the "partial-time, partial-space" heating mode:

(1) To have the function of room-specific regulation and ON as-needed, allow independent control over different rooms, meet the needs of intermittent heating, and make full use of farmers' behavior energy saving to achieve maximum efficient heating;
(2) To keep the temperature settable to meet the requirements for different functional rooms at different periods;
(3) To have the advantages of quick start and stop, fast heating, etc., thus reducing ineffective thermal operation;
(4) To keep the equipment installed independently, without additional terminals, featuring easy installation and maintenance;
(5) To enable one-click operation, easy start and stop, and meet the needs of different groups.

1. Flexible Regulation of the Usage Pattern of Heating Equipment

For users with the flexible regulation and room-specific installation of air-to-air heat pumps in 2.4.1, Tsinghua University conducted a long-period test on actual rural residential buildings in different regions. Figure 3.18 shows the overall heating demand of each room of an RRB in northern China for three years. Each continuous crimson line in the figure indicates the "ON" of heating equipment, i.e., with heating demand. It can be seen that each room is intermittently heated, and there are obvious differences in the heating period and duration of different rooms.

3.4 Analysis of Heating Modes and Energy-Saving Paths of Northern RRBs

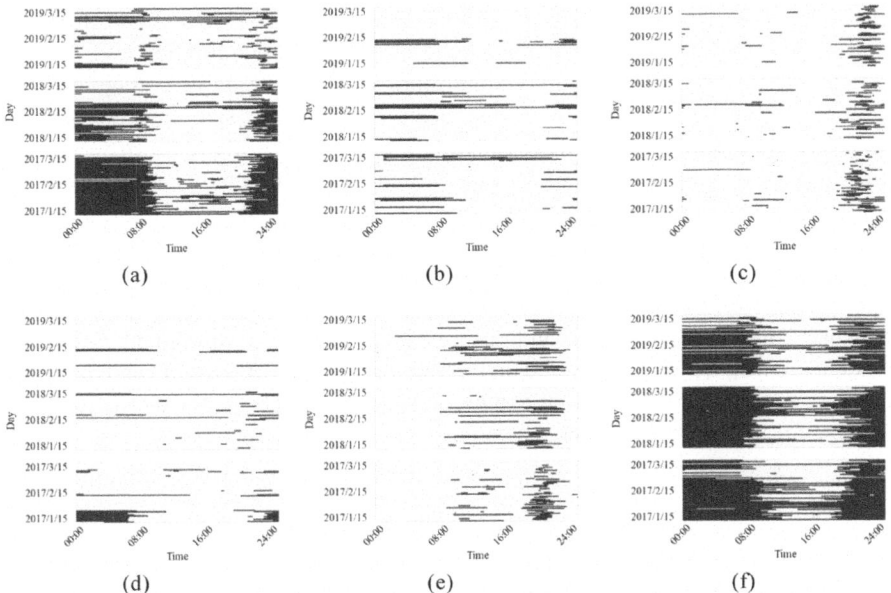

Fig. 3.18 Heat map of 3-year heating demand distribution of air-to-air heat pumps in each room of an actual RRB in northern China. **a** East bedroom; **b** Middle bedroom; **c** Bathroom; **d** West bedroom; **e** Kitchen and dining room; **f** Living room

The east bedroom (i.e., master bedroom) and the living room (where most of the users' activities take place) show the highest utilization of air-to-air heat pumps and the corresponding heating demand. The air-to-air heat pump in the living room is kept on in most cases, except at noon when the outdoor temperature rises. As can be seen from Fig. 3.18a, the east bedroom has the largest difference in three-year usage rate. The return visit showed that this was because the heating demand of the east bedroom gradually decreased as young children grew up. Users often chose continuous heating during the night sleep period when children were younger in the early stage, and the gradually declined frequency of use over the next two years was ascribed to the timing off and otherwise of the air-to-air heat pump after falling asleep. In addition, the bathroom and kitchen have heating demands for a short time in the morning and evening. This is in line with the function of the two rooms and the type of indoor activities. The west bedroom and the middle bedroom are randomly heated based on the availability of guests. This way of heating is completely controlled by the users.

Over the whole heating period, the uneven distribution of heating demand in the rooms is evident. The distribution characteristics of heating demand can be explained by the difference in rooms and in time, especially in room size and heating duration. The heating load of rooms is mainly determined by the building envelope, room size, and heating duration. The first two depend on the building types, and the heating duration on the users' personalized heating demand. Given their functions, different rooms often correspond to varying in-room hours and heating duration.

Table 3.10 shows the statistics on the room size and the daily average heating duration. It can be seen that the rooms are not uniform in size, and the heating duration varies greatly. Known for maximum demand, the living room features a daily average heating duration of 15.6 h, accounting for more than half of the entire heating period, followed by the east bedroom, kitchen, and bathroom. The daily average heating duration for the two guest bedrooms is less than 3 h.

The heating demand of the above RRB shows obvious characteristics of "partial-time, partial-space," which is fundamentally different from the traditional "full-time, full-space" heating mode of urban housing. Only the heating equipment with independent control, flexible regulation, and rapid response in each room can meet such heating demand to the maximum extent, and reduce excessive heating as much as possible for the purpose of energy saving.

2. The Influence of Heating Equipment with Different Degrees of Regulation on Heating Mode and Energy Consumption

To explore the influence of heating equipment with different degrees of regulation on heating mode and energy consumption, and eliminate the user differences, two common heating systems on the market are selected to test the operating mode of the same household under different heating systems.

It is a common inverted "U" shaped RRB in northern China, as shown in Fig. 3.19.

Before the test, the RRB was renovated for clean heating, ending up with an air-to-water heat pump and an underfloor heating system for the north house heating. The underfloor heating system is available in four rooms of the north house: guest bedroom 2, master bedroom, living room, and guest bedroom 1. To compare the operation of different heating systems, another set of low-ambient temperature air-to-air heat pump heating system with room-specific heating is installed (Fig. 3.20), covering all rooms of the existing underfloor heating system. The two systems are not used simultaneously. The operating state information of the water heater and air-to-air heat pump is obtained by running power.

The household was heated by the air-to-air heat pump from November 15, 2022, to January 15, 2023, and by the air-to-water heat pump from January 16, 2023, to

Table 3.10 Room size and daily average heating duration of an actual RRB in northern China

Room	Living room	East bedroom	Kitchen and dining room	Bathroom	Middle bedroom	West bedroom	Total
Area (m^2)	35.6	15.5	25.2	12.7	16.8	15.5	121.3
Daily average heating duration (h)	15.6	6.6	2.2	1.8	2.2	1.0	–
Proportion of heating duration (%)	65.0	27.5	9.2	7.7	9.2	4.2	–

3.4 Analysis of Heating Modes and Energy-Saving Paths of Northern RRBs

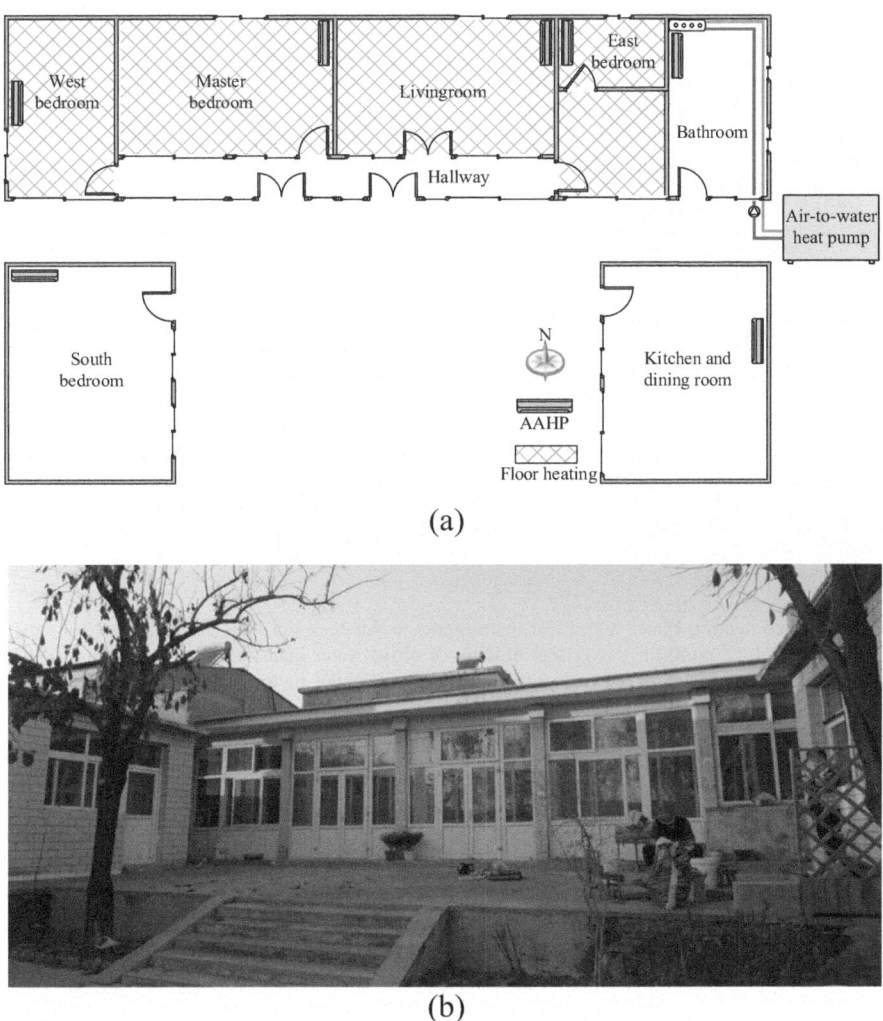

Fig. 3.19 Floor plan of RRB for comparison testing of heating equipment **a** Floor plan; **b** RRB photo

March 15, 2023. Figures 3.21 and 3.22 show the running power of the air-to-air heat pump and air-to-water heat pump in each room during the heating period. It can be seen that the air-to-air heat pump was not turned on continuously at the same time, while the air-to-water heat pump had been running continuously.

Figures 3.23 and 3.24 show the ON–OFF status of the heating equipment in each room, hourly power-on probability, and room temperature when the air-to-air heat pump and air-to-water heat pump were operating. It can be seen that the mean room temperature of the two heating systems is above 19 °C, which can meet the heating

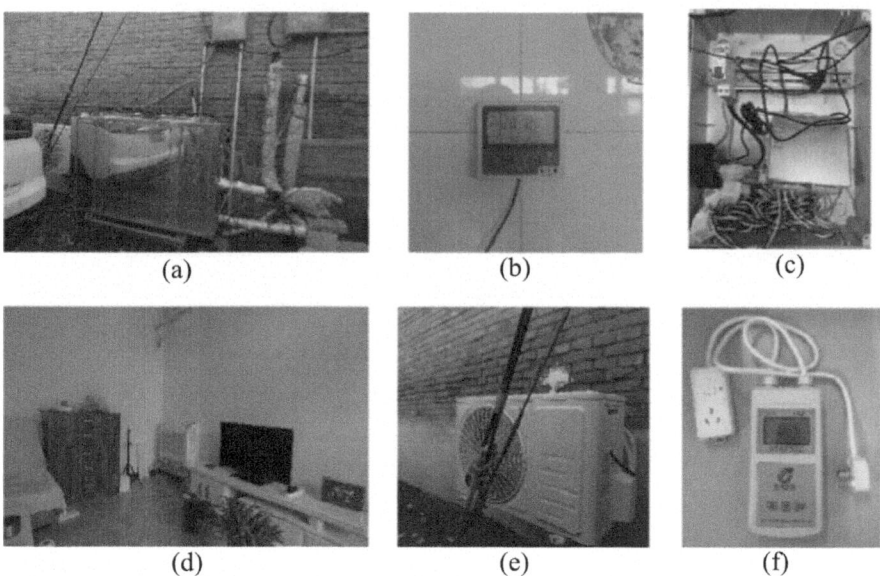

Fig. 3.20 Two heating systems and test instruments. **a** Air-to-water heat pump; **b** Control panel of air-to-water heat pump; **c** Power test module of air-to-water heat pump; **d** Indoor unit of the air-to-air heat pump in the master bedroom; **e** Outdoor unit of the air-to-air heat pump; **f** Test power meter of the air-to-air heat pump

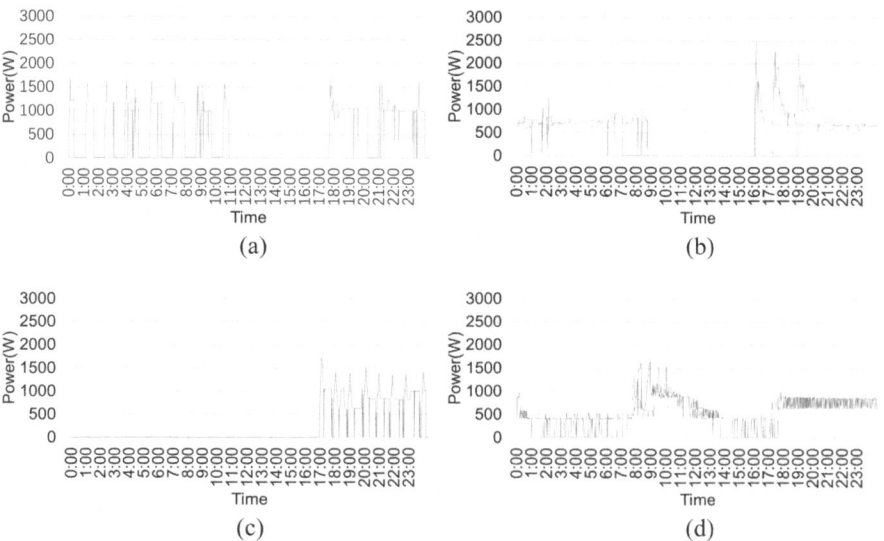

Fig. 3.21 Running power of the air-to-air heat pump in each room*. **a** Guest bedroom 1; **b** Living room; **c** Master bedroom; **d** Guest bedroom 2. *Note* The color version of this figure can be viewed by scanning the QR code in the Table of Contents

3.4 Analysis of Heating Modes and Energy-Saving Paths of Northern RRBs

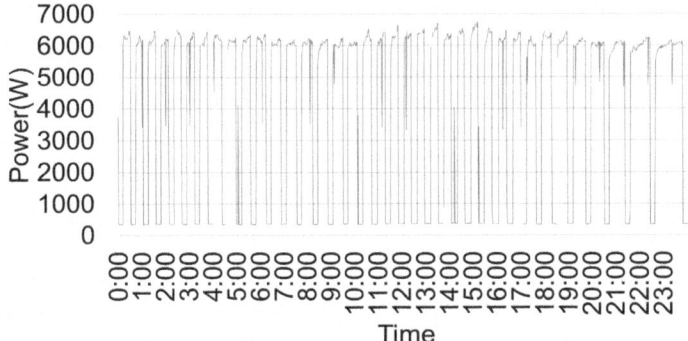

Fig. 3.22 Running power of the air-to-water heat pump

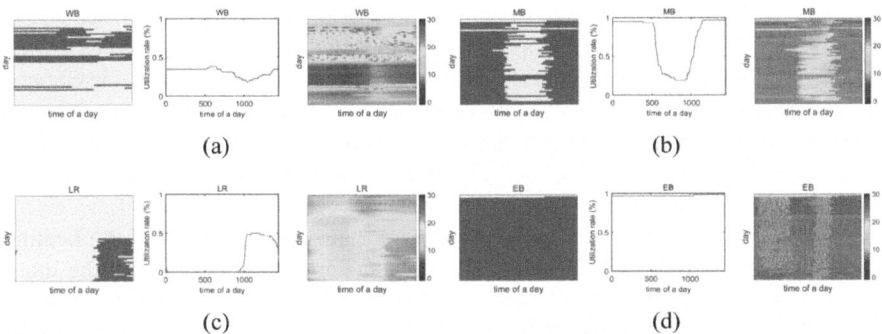

Fig. 3.23 Heating diagram of air-to-air heat pump, hourly heating probability, and room temperature (*Note* Blue represents the heating period)*. **a** Guest bedroom 2; **b** Master bedroom; **c** Living room; **d** Guest bedroom 1

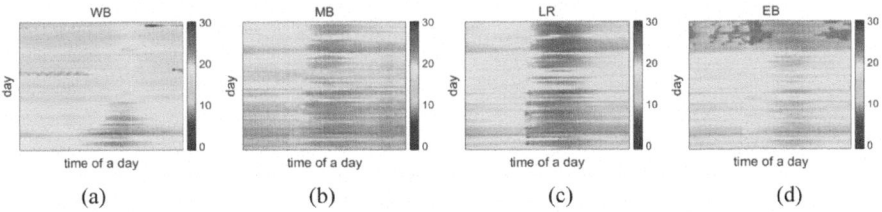

Fig. 3.24 Daily room temperature in each room during use of air-to-water heat pump*. **a** Guest bedroom 2; **b** Master bedroom; **c** Living room; **d** Guest bedroom 1

demand of users. The room temperature is more stable during the use of the air-to-water heat pump, but lower than that during the startup period of the air-to-air heat pump.

To eliminate the impact of room temperature differences, the same room temperature was set according to the actual operating period of the heating equipment in each room, and the energy consumption was simulated using DeST. According to the operation rules of the air-to-air heat pump, the simulated heat load in the heating season is 2,245 kWh. Based upon the continuous operation of the air-to-water heat pump, the simulated heat load in the heating season is 3,361 kWh, 1.5 times that of intermittent operation.

The above test results show that the heating demand of users is characterized by "partial-time, partial-space". For the heating system of "partial-time, partial-space" at the end of convection with room-specific installation and quick start and stop, users can express the heating demand by regulating the equipment, thus realizing on-demand control and energy saving. In the case of a "full-time, full-space" heating system, users cannot achieve on-demand heating, subject to the system regulation capacity. Therefore, there are significant differences in heating load between the two systems. In other words, when the heating demand is unchanged, allowing for the regulation mode, system response speed, adjustable degree, and system thermal inertia under different heating equipment, users will shift their regulatory behaviors, resulting in different equipment operating modes and heat supply modes. Heating equipment will have limitations and influences on the heating behavior of users, and the participation of users in regulation are closely related to the thermal inertia and regulation characteristics at the system end. The expression of user needs is somewhat restricted under different heating systems and regulatory mechanisms, thus showing different actual operation rules.

In a detached building such as RRB, users are also the regulator of the heating system. The heat consumption and operating costs are closely related to the interests of users, and users are more actively involved in energy conservation. In the heating system where each room can be regulated separately and heated rapidly, users' regulation capability is fully mobilized and exerted. Meanwhile, users can guarantee their personalized demands in different rooms and different periods, thus ensuring the expression of their willingness to adjust to demand, and maximizing the coincidence and matching of supply and demand. Air-to-air heat pumps with flexible regulation and room-specific installation can meet the heating demand of users, and reduce ineffective heat supply and minimize energy consumption.

3.5 Summary

First of all, this chapter defines the relevant concepts of rural areas in China, calculates the current situation and main characteristics of energy consumption in rural construction and agricultural production, summarizes the progress, effects, and problems of the clean heating campaign in northern rural areas, and analyzes the heating mode and actual demand of RRBs in northern China. It should be noted that at present, China's rural areas are in a special period of rapid transformation, and the energy consumption data has been constantly updated and changing. This chapter gives only the calculation results of the section at a specific time. These results can

3.5 Summary

still provide some reference for the quantitative understanding and analysis of the status quo, trends, and problems of rural energy use in China.

① In the commercial energy consumption of rural buildings in 2022, the consumption of commercial coal, LPG, electricity, and natural gas was 95 million tons (67 million tce), 8.07 million tons (14 million tce), 348.1 billion kWh (105 million tce), and 12.4 billion m^3 (15 million tce) respectively, totaling 201 million tce; the total consumption of non-commercial energy, i.e., biomass (including firewood and straw), was 94 million tons (52 million tce). Commercial energy and non-commercial energy account for 79.4% and 20.6%, respectively.

② Since the launch of the *Air Pollution Prevention and Control Action Plan* in 2013, the consumption of bulk coal in rural areas has decreased from 197 million tons in 2014 to 95 million tons in 2022, down by about 51.8%, thanks to the Three-Year Action Plan for protection of the Blue Sky, especially the implementation of the clean heating pilot action in northern China. The Beijing-Tianjin-Hebei region and its surrounding areas account for 49% and 82% of China's $PM_{2.5}$ and CO_2 emission reductions caused by domestic energy consumption in rural areas, respectively.

③ In 2022, the total biomass consumption and its proportion in household energy consumption in rural areas of China continued to decrease. Compared with 2014, the biomass energy based on straw and firewood decreased by about 87 million tons, the proportion of which fell from 31.2% to 20.6%.

④ Considering the building form, occupant type, and income level, as well as the actual building space usage pattern, thermal comfort characteristics, and heating demand of RRB, the heating system and equipment of "partial-time, partial-space", featuring room-specific installation and flexible regulation, can better match the actual demand characteristics of RRB in northern China. By comparison, the menu-based "targeted" renovation scheme of building envelop in RRB can achieve a reasonable energy-saving effect, and even optimize the initial investment in renovation. Thus, it solves the game problem of scheme selection, energy-saving effect, and investment cost.

⑤ In the past five years, remarkable achievements have been made in the introduction of policies, the number of clean energy alternatives, regional air quality, and human health improvement for clean heating in northern China. However, further improvements or enhancements are required at the three levels of technology implementation, policy support, and sustainable development, including guiding overall policy planning and management, scientific formulation of the technical route, synchronous promotion of energy conservation in RRB building envelope, equipment and operation subsidy methods, rural energy use and low-carbon development issues, and promoting the application of appropriate models.

Open Access This chapter is licensed under the terms of the Creative Commons Attribution-NonCommercial-NoDerivatives 4.0 International License (http://creativecommons.org/licenses/by-nc-nd/4.0/), which permits any noncommercial use, sharing, distribution and reproduction in any medium or format, as long as you give appropriate credit to the original author(s) and the source, provide a link to the Creative Commons license and indicate if you modified the licensed material. You do not have permission under this license to share adapted material derived from this chapter or parts of it.

The images or other third party material in this chapter are included in the chapter's Creative Commons license, unless indicated otherwise in a credit line to the material. If material is not included in the chapter's Creative Commons license and your intended use is not permitted by statutory regulation or exceeds the permitted use, you will need to obtain permission directly from the copyright holder.

Chapter 4
Key Solution to China's Energy Issue: Energy Revolution Initialized in Rural Areas

Achieving the carbon peaking and carbon neutrality goals is an important milestone in the energy revolution. In general, the energy system must undergo a transition from the current reliance on fossil fuels to zero-carbon energy in order to achieve carbon neutrality. The central government clearly calls for "establishing the new before abolishing the old", and "achieving energy transition under the premise of ensuring social development and economic growth". Therefore, it is necessary to define the structure of the future zero-carbon energy system, set the ultimate goal of energy transition, and on this basis, chart a transition path and make clear priorities.

The energy system involves the whole process of energy production, energy conversion and transmission, and energy end-use. The ultimate goal of the energy system is to provide a reliable energy supply for final energy consumption. At present, energy end-use is usually divided by nature into three areas: industry, construction, and transportation. Industrial production is subdivided into process industry and non-process industry by the characteristics of energy use in the production process. Metallurgy, nonferrous metals, building materials, and chemical engineering are process industries. Fossil fuels enter parts of the production process both as fuel and as feedstock. These production processes need the radical changes to achieve carbon neutrality. Many industries such as machinery industry, electronics, and light industry are non-process industries, where the energy consumption of the production process is merely electricity, heat, and fuel. The characteristics of energy use in construction and transportation are very similar to those in the non-process industry. The energy sources for building operations are electricity, heat, and fuel, while the transportation service is mainly powered by electricity and fuel. Since the burning of fossil fuels inevitably releases carbon dioxide, the zero-carbon fuels in the future energy system can only be derived from biomass fuels or hydrogen fuels produced from zero-carbon electricity by electrolysis of water. Therefore, it is necessary to minimize the demand for fuel in non-process industry, building, and transportation, to replace the demand for fuel with electricity or central heating, and to achieve full electrification in all three fields as far as possible.

The primary objective of China's energy revolution is to build a zero-carbon power system and a zero-carbon fuel production, supply, and marketing infrastructure. Rural areas, endowed with abundant space and biomass resources, have the potential to emerge as significant suppliers of zero-carbon electricity and zero-carbon fuels derived from biomass materials. Rural areas will change from energy consumers to energy suppliers in the future zero-carbon energy framework. The rural new energy system will be a key part of the energy transition in China.

4.1 Transition to Zero-Carbon in China's Electricity Sector

In 2021, China's energy-related carbon dioxide emissions were about 10 billion tons, of which carbon emissions from power generation were about 4 billion tons. The electric power sector is the dominant carbon emission sector in China, and will undertake the transfer of carbon emissions from other industries in the future. This sector shoulders the toughest task of reducing emissions. In 2021, the total power consumption of the whole society was 8.5 PWh. It is estimated that by 2050, the total power demand of the whole society will reach 13 PWh, excluding the power consumption of heating systems (Table 4.1).

To meet this total electricity demand, the power supply structure needs to be continuously optimized, and the proportion of new energy installed capacity and energy production continues to increase (Table 4.2). The total installed capacity of the power system is expected to reach 8.8 billion kW in 2050, and the proportion of non-fossil energy installed capacity will increase from 46% in 2021 to 92% in 2050. The share of non-fossil energy production will increase from 34% in 2020 to 90% in 2050. Taking into account the future demand for zero-carbon fuels like electrolytic hydrogen production and partial curtailment of wind and solar power, the annual energy production is expected to reach 14.5 PWh in 2050.

In order to prevent thermal pollution of offshore seawater caused by nuclear power heat discharge, a comprehensive plan should be developed for coastal sites suitable for nuclear power plants, considering their exclusive location on the coast. China's future nuclear power installed capacity will be limited to 200 million kW, providing

Table 4.1 National total power consumption and future forecast (unit: PWh)

Consumption field	2021	2050
Industry	5.5	6.5
Building operation	2	4
Transportation	0.5	2
Others	0.5	0.5
Total	8.5	13

Note The 2050 figures do not include electricity required for heat supply

4.1 Transition to Zero-Carbon in China's Electricity Sector

Table 4.2 China's power supply structure and annual energy production forecast

Power supply structure	2021		2050	
	Installed capacity (100 million kW)	Annual energy production (PWh)	Installed capacity (100 million kW)	Annual energy production (PWh)
Nuclear power	0.5	0.4	2	1.5
Hydropower	4	1.5	5	2
Wind and solar power	7	1	74	9.5
Thermal power	13	5.6	7	1.5
Total	24.5	8.5	88	14.5

an annual electricity supply of 1.5 PWh throughout the year. Hydropower serves as a flexible and adjustable source of zero-carbon electricity. With over 90% utilization rate in China's hydropower resources except for the Yarlung Zangbo River basin in Xizang, its installed capacity reached 400 million kW in 2021, with an annual energy production of 1.5 PWh. Considering climatic conditions and geographical factors, China's installed hydropower capacity will be capped at 500 million kW, with an available energy production of 2 PWh throughout the year. Based on the total energy production of 14.5 PWh, there remains a power gap of approximately 9 PWh, which can be supplemented by wind and solar power. According to the average annual generating hours of 1,300 h, the combined installed capacity of wind and solar power will be 7.4 billion kW. China's total installed capacity of wind and solar power was 740 million kW in 2022. This implies that the total installed capacity of wind and solar power will increase to tenfold compared to the current level.

Unlike thermal power or nuclear power, wind power and solar power are low-density resources. Since it relies on natural solar radiation and wind, there must be enough space to generate the required power. If estimated at 100 W/m² of generated power, the 7.4 billion kW wind and solar power requires 123 million *mu* (82,000 km²) of land space. The development of wind and solar power is first challenged by how to determine the installation space.

Northwest China has vast expanse of the Gobi Desert, with good solar radiation intensity and wind conditions. Should we prioritize the development of wind and solar power in these areas? China's geographical resources and environmental conditions have resulted in more than 70% of the population being concentrated in the central and eastern regions of the country. Consequently, this concentration also leads to a centralized demand for electricity in these areas. If the main power supply comes from the northwest region, it is essential to continue working on the "West-to-East Electricity Transmission" Project, aiming to transmit the wind and solar power from the northwest region to the central and eastern areas with intensive power load by using the long-distance, high-power transmission lines. There are three problems with this: First, with the declining cost of wind and solar power generation installations, the cost of long-distance transmission facilities has been comparable to those of wind and solar power generation facilities. Therefore, deploying wind and solar power in the northwest region and relying on the "West-to-East Electricity Transmission" Project will significantly increase the cost of electric power construction. Second, due to the geographical constraints of the Hexi Corridor, the construction and spatial layout of transmission channels are significantly limited, and cannot accommodate more than 4 billion kW of transmission lines. Third, the diurnal variation of wind and solar power is somewhat similar to the changes in terminal electrical load. Long-distance transmission has to be regulated through energy storage on the power side. On the user side, energy storage measures are also utilized to synchronize changes with power demand. The two regulatory and storage initiatives will also lead to a substantial rise in the cost of constructing electric power infrastructure. Therefore, considering the objective conditions such as space and resources in the western region, priority should be given to implementing a local balancing system of distributed generation to meet the increasing power demand in the central and eastern regions. This approach is economically viable and optimizes resource utilization. The distribution of China's installed wind and solar power capacity in the eastern and western regions should be in the range of 7:3 to 6:4, meaning that 60% to 70% of the wind and solar power should be concentrated in the central and eastern regions with high electricity demand.

The central and eastern load-intensive areas will develop 5 billion kW of wind and solar power. Where can we find enough space for installation? Tsinghua University, in collaboration with the Land Satellite Remote Sensing Application Center under the Ministry of Natural Resources, has obtained statistics on the roof area of urban and rural buildings in China by utilizing high-resolution satellite images. According to conservative analysis and estimation, urban roofs can accomodate 870 million kW of photovoltaic (PV) panels, generating an annual energy production of 1 PWh. Rural building roofs, including warehouses and other non-residential buildings, can support 2 billion kW of PV panels, producing an annual energy output of 2.9 PWh. In this way, urban and rural rooftop PV systems can provide 1/3 of the installed capacity and energy production of wind and solar power needed by the country in the future. And 75% of such rooftop PV systems are distributed in the central and eastern regions. There are already 2 billion kW of rooftop PV installed resources. An additional 1.3 billion kW of concentrated and distributed wind and solar power in the central and eastern regions will be sufficient to address the space requirements for the development of wind and solar power.

4.2 Alternatives to Fossil Fuels and Supply of Zero-Carbon Fuels

Zero-carbon energy is poised to dominate the future energy landscape. According to relevant forecasts, China will need 1.45 billion tce of fuels for power generation, industrial production, and transportation in 2060. As a zero-carbon fuel produced in nature, biomass energy should be the fuel of choice in the future. Analyses show that the biomass resources of agriculture, animal husbandry, and forestry in China have the potential to provide about 700 million tce per year, excluding a small amount of demand for feed, raw materials for agriculture, and by-products. The urban restaurant waste, green waste, and industrial processing outputs have the potential to provide about 300 million tce. This makes it possible to provide 1 billion tce of fuels from biomass fuels. The remaining 450 million tce will be either by burning coal and gas, with recycled CO_2 captured in its flue gas emissions through CCS (carbon capture and storage), or by producing hydrogen through water electrolysis, and then synthesizing hydrogen into zero-carbon fuels. However, both the fossil energy combustion with CCS and the synthetic fuel of hydrogen incur combined costs much higher than biomass fuels. Therefore, the primary task of developing zero-carbon fuels should be to fully utilize the biomass resources generated by agriculture, animal husbandry, and forestry, and convert them into commercial fuels. This can provide low-cost, zero-carbon fuels for a zero-carbon society, and enable farmers to gain considerable new economic benefits.

Biomass resources are meant to be consumed and utilized. Straw, manure, and restaurant waste will be generated as a result of necessary social activities like planting, breeding, and catering. There is no alternative but disposal. Although biomass is generally considered to be zero-carbon, different forms of consumption produce nitrous oxide and methane. Methane and nitrous oxide are the second and third contributors to greenhouse gas in the atmosphere, with global warming potentials[1] about 28 and 300 times that of carbon dioxide, respectively. Since September 2020, China has indicated on several major international occasions that it will strengthen the control of non-carbon dioxide greenhouse gases such as methane. In November 2023, multiple departments jointly issued the *Action Plan for Methane Emission Control* to set requirements for methane control. On the whole, straw returning and discarding are not the best solutions, as detailed in the *Annual Report on China Building Energy Efficiency 2020 (Rural Residential Building)*. Looking at the rationalization of biomass in terms of carbon neutrality, it shall be optimized for energy efficiency to achieve dual benefits. In short, biomass fuel will be the only zero-carbon fuel available in large quantities in the future, and should be considered and developed as a strategic resource.

4.3 Rural Areas to Play a Dominant Role in the Future Energy Structure System

Space is the most crucial resource for the development of zero-carbon energy sources like wind, solar, and biomass. Rural areas offer plenty of space for such developments (see in this chapter). At present, rural areas consume a total of less than 350 billion kWh of power, and only 500 billion kWh when combined with agricultural

---

[1] The value of the relative radiation effect of a given substance compared to carbon dioxide over a certain time integral range.

production. In other words, the actual power consumption in the rural areas is only 1/5 of the potential potential of rooftop PV power generation in those rural areas. Therefore, fully exploiting the rural rooftop space resource for PV power generation will provide the energy required for living, production, and transportation, and the electricity generated will surpass that from coal, oil, gas, and biomass, ultimately achieving full electrification in the rural areas. All fuel coal, oil, and gas fuel will be completely eliminated in agricultural, forestry, and pastoral areas, and there will be no more burning of straw. Full electrification may bring back the long-lost blue sky and white clouds in rural areas. Given the full electrification of rural areas in the future, power consumption will increase with the improved living and production standards. Despite this, rural areas can still export 1–1.5 PWh of electricity to the large grid to support the national energy system.

Practice shows that the roof of a residential rooftop building in northern China can install a PV capacity of 20–40 kW, with an annual energy production ranging from 25,000 to 50,000 kWh. At present, most rural PV systems are connected to the grid through direct inverter feed-in, which is subject to the transformer distribution capacity of 5–6 kW per rural household. The actual PV installed capacity in rural areas accounts for approximately 20% of the total PV installed potential, while 80% of the rooftop PV installed potential is wasted. With scientific and reasonable design and the use of a new PV power system featuring "photovoltaic, energy storage, direct current, and flexibility (PEDF)", rural areas will be transformed from energy consumers into energy prosumers, playing a key role in China's future energy structure system (see Chap. 5). Indoor DC distribution and direct PV access can fully meet the energy demand for cooking, heating, lighting, and home appliances. The charging points can also charge electric vehicles and motor farm implements to meet the energy use requirements for production and transportation. When the battery pack in the battery swapping mode is used to power vehicles and farm machinery, or the two-way charge and discharge mode is used, each household has a storage capacity of more than 60 kWh. This capacity can ensure normal energy supply for daily life even during three consecutive cloudy days. The excess power from each household is aggregated into the village-level DC microgrid to supply energy to public lighting, production, and other electricity facilities. And surplus power is then transmitted to the grid by inverter step-up process. At this time, individual households and the public can utilize electricity storage capacity to store excess power and provide a consistent electricity supply over an extended period. Since the daily peak of PV on sunny days is only 3–4 h, the storage regulation can extend the time of sending electricity to the grid through the transformer to 12–15 h per day, even full power supply in the evening peak from 17:00 to 20:00. It helps the grid deal with power shortages at evening peak. The DC/AC inverter connected to each distribution transformer can adjust the DC busbar voltage according to the output power: Where the on-grid power is to be increased, it reduces the busbar voltage, and the DC/DC converter connected between the indoor DC system and the off-grid of each household will increase the power transmission into the village-level microgrid, thereby improving the on-grid power of the distribution transformer region. When the DC/AC inverter detects a higher on-grid power, it can increase the busbar voltage, and thus reduce the output power of each household. The distributed coordinated control based on DC busbar voltage can realize the safe operation of the power system in the village, and make full use of the distribution capacity of the rural power grid for power transmission. In other words, the "PEDF" new power system can realize 100% development of PV potential in rural areas, without increasing the cost of transformer expansion. This basically achieves zero-carbon emissions in the rural energy system. Since the time of on-grid power supply is decoupled from solar radiation through power storage, the on-grid power supply will be transformed from PV "garbage electricity" to high-quality electricity for grid peak shaving. At this time, rural areas can obtain higher feed-in tariffs according to the actual demand of the power grid.

To achieve the above goals, it is preliminarily estimated that each farmer household should have an average of 50–60 kWh of electricity storage capacity. At present, the cost of lithium batteries continues to decrease, and the minimum cost of 50 kWh lithium batteries is reduced to RMB 30,000. This should be the main content of home appliances going to the countryside. Every household is equipped with energy storage resources, laying the foundation for complete electrification and the full development of rooftop PV resources. In fact, the farmer households may have access to indirect energy storage resources through full electrification. To develop electrification of farm machinery, all kinds of mobile farm machinery are powered by batteries. These batteries also serve as excellent energy storage resources. The use of motor farm implements is much lower than that of private cars. This allows the development of a standard modular battery-swapping mode. Each farm machinery owner can stock up on batteries, which can be matched to a variety of farm machinery. In this way, they can significantly decrease the total installed capacity and the installation cost of electric motor farm implements. Villages that have fully electrified their farm machinery can achieve 60 kWh of electricity storage capacity per household. One electric vehicle per household can achieve an electricity storage capacity of 50 kWh. In addition, both the agricultural production and the processing process of agricultural and by-products can realize the demand response mode for electricity consumption. Therefore, rural areas should fully utilize these flexible resources, conveniently consume PV energy, and feed in the remaining electricity in an orderly manner.

When the rural areas are fully electrified, the substituted agricultural straws, forestry branches, animal husbandry manure, and other biomass materials can be processed into the solid, liquid, or gaseous form of commercial fuels, and then enter the fuel market. At present, corn stalks and fruit branches can be processed into pellet fuels, which can be burned cleanly in specialized stoves. Wheat straw and rice straw can be compressed into a compact with a relative density of 1.2, which can be burned in boilers as an alternative to coal fuel. The combustion efficiency of biomass solid molding fuels can exceed 65%. This reduces the pollutants emitted during the combustion, and greatly improves the energy efficiency. The largest processing cost of briquette fuel is electricity consumption, and PV can provide sufficient zero-carbon electricity. The production process of biomass compression process becomes intermittent power consumption, operating from sunrise to sunset to alleviate the strain on electricity storage. Wet biomass, such as livestock manure, can be used to produce biogas in large biogas digesters. Carbon dioxide and other components are then separated from the biogas, and methane is further purified to become high-quality gas in the gas market. As a result, rural areas of the future will export food, as well as valuable zero-carbon electricity and zero-carbon fuels.

Why is biomass fuel not preferred for self-use but processed into commercial fuels for export? A product should be storable and easily transportable to enter the commodity market. Electricity is difficult to store and transmit; it should be used as preferentially as possible and consumed on-site. Solid and gaseous biomass fuels have characteristics similar to coal and should be prioritized for sale.

4.4 Contribution of Initiating Energy Revolution in Rural Areas to Carbon Neutrality in China

To pave the way for an energy revolution and achieve carbon neutrality in rural areas, it is essential to decrease energy intensity and enhance energy efficiency from the demand side. Subsequently, on the supply side, there is a need to actively promote the development of renewable energy, focusing on rooftop PV and biomass energy. This section focuses on analyzing various different future rural development scenarios based on supply and demand balance. The statistical analysis of rural PV and biomass resources in China is detailed in this chapter.

As the urbanization rate continues to increase in the future, the rural population will likely continue to decline. Assuming an 80% urbanization rate by 2060, the rural population will be about 260 million. Based on the current and future development trends in China's rural areas, we can anticipate future changes in rural energy demand. Table 4.3 shows the predetermined values of key factors influencing rural energy consumption and production under different development scenarios. According to the *China Statistical Yearbook 2022*, the ownership of certain electrical appliances, such as color TV sets, in rural areas is 116.3 units per 100 households, which has essentially reached its maximum growth potential. There are still discrepancies between the numbers of other electrical appliances and the city. For example, the numbers of washing machines and refrigerators are 96.1 units and 103.5 units per 100 households. The figures will continue to grow and eventually reach the upper limit of the rural economic level. With the improvement of living standards, the ownership of electric vehicles and motor farm implements in rural areas, which are currently limited, will also continue to increase. By 2030, the car ownership per 1,000 people in rural areas is expected to reach 160, according to the China Association of Automobile Manufacturers. In March 2023, the penetration rate of battery electric passenger cars in county and township areas was 16%, while that of plug-in hybrid passenger cars in county and township areas was 8%, as reported by the China Passenger Car Association. In the future, the adoption of new energy vehicles is expected to further increase in rural areas. Motorized farm implements have shown advantages in efficiency, energy conservation, environmental protection, and other aspects. Their

4.4 Contribution of Initiating Energy Revolution in Rural Areas to Carbon ...

Table 4.3 Predetermined values of key factors influencing rural energy consumption and production under different development scenarios

	Scenario	2030 (%)	2040 (%)	2050 (%)	2060 (%)
Energy-saving renovation of building envelope	Base scenario	2	5	7	10
	Carbon–neutral scenario	21	44	67	90
	Scenario	2030 2030 (%)	2040 2040 (%)	2050 2050 (%)	2060 2060 (%)
Proportion of zero-carbon heating users	Base scenario	7	15	22	30
	Carbon–neutral scenario	47	100	100	100
Electrification rate of cooking	Base scenario	7	15	22	30
	Carbon–neutral scenario	47	100	100	100
Penetration rate of electric vehicles and farm machinery	Base scenario	35	45	65	80
	Carbon–neutral scenario	50	70	80	100
Proportion of renewable energy development	Base scenario	5	10	15	20
	Carbon–neutral scenario	23	49	74	100

prevalence in rural areas is expected to continue to rise in the future. The renewable energy resources in rural areas far exceed the energy demand. In addition to meeting the rural energy needs, the continued development of rooftop PV and biomass energy can also supply zero-carbon energy to urban areas. On one hand, it can generate substitution benefits for carbon emissions, while on the other hand, it can boost farmers' income and create economic value. According to the rural renewable energy landscape, biomass energy is mainly supplied to urban areas. Rooftop PV power generation needs to meet rural energy needs, and the remaining parts are exported in an orderly manner according to the grid dispatching requirements.

In the carbon–neutral scenario, rural areas play a crucial role as a production base for renewable energy by actively developing rural rooftop PV and biomass resources. The demand for rural energy has shifted toward electrification at a faster pace. The rooftop PV power generation meets the rural energy consumption and can even supply electricity to urban areas to address fluctuations in the electrical load through the establishment and flexible management of rural microgrids. Biomass fully leverages the benefits of zero-carbon fuel for energy consumption, making it an significant alternative to coal in industrial production. Rural areas are projected to replace all domestic bulk coal by 2030–2035 and achieve near-zero carbon energy consumption by 2040 (Fig. 4.1). In the meantime, rural areas are expected to significantly increase the electrification rate, ensuring that electricity meets the energy demand for transportation, cooking, heating, and agricultural production. In 2060, rural areas will export 2.3 PWh of zero-carbon electricity and 740 million tce of zero-carbon fuels to cities. It should be emphasized that as the proportion of installed renewable energy capacity increases, coal consumption for power generation fails to align with the

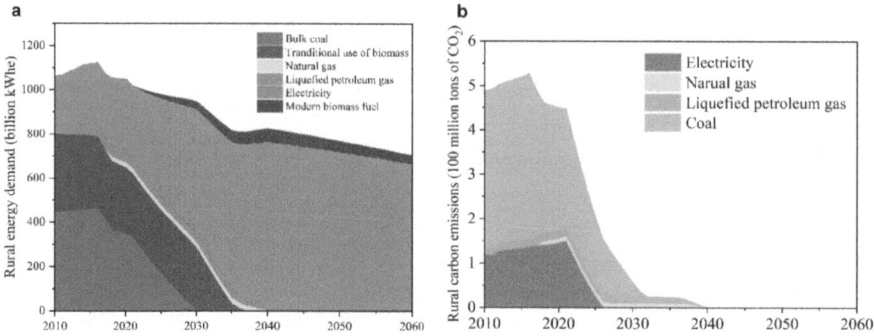

Fig. 4.1 Rural energy demand and carbon emission prediction under carbon–neutral scenario*. **a** Energy demand; **b** Carbon emissions from energy

increasing power consumption trend in rural areas. This book converts non-electric energy sources according to the equivalent electrical method. In the future, as the rural population decreases and building energy efficiency improves, along with changes in the energy structure, the total energy consumption demand for rural residential buildings is expected be lower than the current (2020) total energy consumption.

In the base scenario, rural areas continue to be energy consumers. Farmers in northern China rely on traditional methods like coal to gas for heating, while some use bulk coal and traditional biomass for cooking in rural areas. For the renewable energy supply side, only a small number of rural rooftop PV and biomass resources are developed, while the rest heavily rely on external energy imports. In rural areas, however, domestic bulk coal and loose biomass energy could be eliminated by around 2050. Nevertheless, households in these areas still rely on fossil energy like natural gas and LPG to meet their energy needs, as illustrated in Fig. 4.2. While total carbon emissions in rural areas will be regulated, achieving carbon neutrality is not projected. It will still release approximately 63 million tons of CO_2 by 2060.

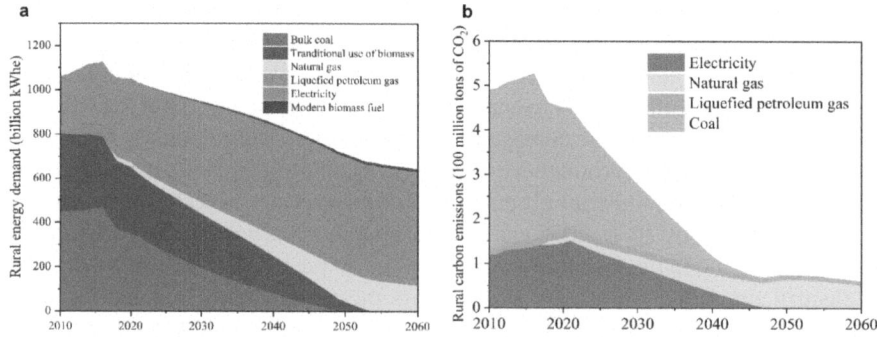

Fig. 4.2 Rural energy demand and carbon emission prediction under base scenario*. **a** Energy demand; **b** Carbon emissions from energy

4.5 As the Breakthrough Point of China's Energy Revolution, Rural Areas are Set to Lead the Way

China's land resources determine that there must be 300 million people living in agricultural, animal husbandry, and forest areas to provide the basic food supplies for the entire population. With a quarter of the population providing only 10% of the food consumed in modern society, it is evident that the average income of the farming, animal husbandry, and forestry population is only 10% of that of the population. This accounts for the significant disparity between urban and rural areas in China. The fragmentation of land makes it difficult for us to improve agricultural production efficiency and eliminate the gap between urban and rural areas through large-scale agricultural mechanization, as seen in the United States. The low labor productivity in agricultural production on scattered land is the main cause of the economic disparity between urban and rural areas in China. To narrow this gap, we must explore new methods of production in rural areas so that farmers can generate additional economic income. The utilization of rural space resources to supply food and energy for society and generate new economic opportunities for farmers can offer a novel approach to bridging the gap between urban and rural areas.

Imagine a grain farmer with centralized land in the future. If he sells grain for RMB 50,000 a year, he can generate an additional income of RMB 15,000 to 20,000 by comprehensively recycling the straw from food crops and processing it into biomass fuels through PV power for sales. Additionally, he can earn RMB 10,000 from selling his surplus PV surplus electricity of 20,000 kWh. His annual income can be increased from RMB 50,000 to RMB 80,000–90,000. This will significantly improve the economic conditions of grain farmers and narrow the gap between urban and rural areas.

What we now face is the contradiction between unbalanced and inadequate development and the people's ever-growing needs for a better life. China still suffers from an income gap among residents, particularly between urban and rural populations. This has become an urgent problem that needs to be addressed to achieve balanced development. *The Outline of the 14th Five-Year Plan (2021–2025) for National Economic and Social Development and Long-Range Objectives through the Year 2035 of the People's Republic of China* clearly calls for "making more notable and substantial progress in achieving common prosperity for everyone". This is based on the new requirements of China to complete the construction of a moderately prosperous society in all respects and to enable people to strive for a better life. It also serves as a new guideline for China to narrow the income gap among residents and foster shared prosperity in the new situation. Under the new situation and new development strategy, narrowing the income gap between urban and rural residents through innovative policy approaches holds significant theoretical and practical importance for achieving fair income distribution, enhancing the quality of economic development, and boosting people's sense of well-being, happiness, and security.

The restructuring of household energy consumption structure in rural areas is not only crucial for controlling environmental pollution sources, enhancing farmers' living standards, and promoting sustainable development, but also essential for achieving China's dual carbon goals. In addition to household energy consumption, agricultural production is an important energy-consuming sector. Despite the widespread adoption of rural mechanization, it is predominantly diesel-powered. Diesel consumption in rural areas accounts for approximately 15% of the country's total annual diesel consumption. The use of bulk coal and diesel fuel has led to significant pollution and carbon emission issues.

To build a new energy system, rural areas must focus on the rooftop PV and DC microgrid. By leveraging low-cost energy storage resources and production, as well as living characteristics in rural areas, the system provides cities with zero-carbon electricity and zero-carbon fuels, while achieving complete electrification. The first practical question before us is where the money for construction comes from. According to the preliminary project demonstration, the average household investment in the rural "PEDF" system is RMB 60,000–100,000 (see Chap. 5 for details). Upon completion, the system can meet all the electricity demands of the rural house, and even export approximately 15,000 kWh of electricity to recoup the initial investment. The construction investment for the public sector is approximately RMB 2–4 million per 100 households, and approximately RMB 10 million of public construction funds are needed for a natural village consisting of 300 households. In the past decade, the state has implemented numerous subsidy policies to address rural energy issues. For example, hundreds of billions of yuan were invested in rural areas in the campaign to promote clean heating in northern China; more than RMB 100 billion was invested in PV poverty alleviation projects during the "13th Five-Year Plan" period; and more than RMB 100 billion was invested in the expansion and upgrading of rural power grids. Besides, there are subsidies available for rural electricity, agricultural implements purchases, diesel, and more. Despite good coverage, these policies and funds are channeled through different channels and do not create synergy. If effectively pooled, these funds can completely solve the problem of investment in rural power grid infrastructure. Upon completion, it will generate stable income for the village collective and villagers, thus forming a positive cycle.

In the future, China is projected to have around 300 million people working in agriculture to guarantee national food security. Only by increasing the economic income of the agricultural population can we truly promote integrated urban and rural development. In addition to boosting agricultural crop yields, we can enhance their income through energy transition. The dual carbon goals proposed by the CPC Central Committee have provided new solutions and new opportunities for rural energy issues. Rural areas can increase their income by at least 50% through energy processing, converting biomass energy, and selling solar power. They may increase their income by about 50% through the development of the tertiary industry and rural revitalization. This is of particular significance for reducing urban–rural disparities.

The construction of a new village-based power system will transform the rural grid from a top-down power supply/distribution network to a distributed generation network. Domestic policy mechanisms are expected to be improved, including the enhancement of the connection between the distributed generation network and the local power grid, the authority for its management and operation, regulations for on-grid electricity sales, and electricity selling prices. Only through experiments and in-depth research on typical model rural power grids can we establish a new mechanism and a new business model that meet the power grid security requirements, support the on-grid sale of electricity, and fulfill the normal requirements of rural electricity consumption. This new model holds significant reference value for the further advancement of urban distributed power systems and active distribution networks. The new power system requires new supporting policy mechanisms, which must be developed through continuous trial and error and iteration in practice. Based on the policy mechanism for constructing the new power system in rural areas, discussions can be regarded as a great place to experiment with the policy mechanism of the new power system. it can play a pioneering role in further realizing the construction of a new power system that integrates both concentration and distribution throughout the country.

The construction of a zero-carbon commercial fuel base based on biomass materials must involve all the stages, including processing, procurement, distribution, and sales, and establish a comprehensive service platform. In fact, the past 20 years have witnessed numerous domestic enterprises engaging in the commercialization of biomass energy. In the absence of suitable platforms and policy mechanisms, most of them failed due to issues with the purchase price of raw materials, the selling price of products, and the identification of the state's subsidy mechanism. Therefore, we must systematically proceed from the healthy development of the entire industry chain, address the future demand for zero-carbon fuels, make comprehensive planning, balance the interests of all stakeholders, and establish an efficient platform to support the development and utilization of commercial fuels derived from biomass materials.

To achieve the establishment of a new urban and rural energy supply system, it is essential to devise a novel policy mechanism that aligns with the characteristics and regulations of the new energy supply system to accommodate its growth. This is the policy mechanism revolution in the "four revolutions" of energy. Such a new policy mechanism needs to be tested through trial and error in practice, and even iterated through engineering practice. Rural areas could be the setting for such pioneering experiments. China's agrarian revolution, reform, and opening up were initially implemented in rural areas and achieved success through the strategy of "Encircling the Cities from the Rural Areas". The revolution of the new energy supply system also needs to be first tried in rural areas, making them the pioneers of policy mechanisms.

4.6 Practical Significance of Rural Energy Revolution

A nation will prosper when its agriculture booms. Comprehensively promoting rural revitalization is an important task to build up China's strength in agriculture in the new era. General Secretary Xi Jinping stressed at the Central Conference on Rural Work that "Efforts should be made to comprehensively promote the revitalization of rural industries, talents, culture, ecology, and organizations. With overall planning and synergy, key areas should be focused on to strengthen the weak links". This important deployment is far-sighted, with rich connotation and clear requirements. It is the fundamental guideline and action guide for doing a good job in "agriculture, rural areas, and farmers" and comprehensively promoting rural revitalization in the new era. Advancing the rural energy revolution is an inherent requirement of ecological civilization construction, an important part of China's energy revolution, and an important measure to solve the issues relating to agriculture, rural areas, and farmers. It is of great significance for promoting the development of beautiful rural areas and implementing the strategy of rural revitalization.

1. Rural Industry Revitalization Backed by Energy Revolution

Industry revitalization stands as the foundational pillar and top priority in rural revitalization. How can we ensure agriculture becomes a profitable venture with a promising prospect, elevating farmers' incomes and rendering farming an appealing career choice, thereby transforming rural areas into a beautiful place to live and work? This is a big topic. According to the survey data, rural residents' expenditure on the commodity energy consumption accounts for more than 6% of the total household income and 11% of the total household expenditure. In 2021, the per capita income of rural residents was RMB 18,931, and the per capita expenditure was RMB 15,916. The annual energy consumption expenditure of 500 million rural residents is about RMB 570–870 billion. With the new power system based on rooftop PV, farmers will most directly benefit from the non-utility power supply and substantial income from electricity sales. This approach effectively unlocks previously untapped resources, such as idle rooftop space. In 2021, China's grain output exceeded 680 million tons, and the yields of legume, tuber, oil, and sugar crops were close to 200 million tons. These agricultural products will produce a lot of straw, residue, and other wastes in the production and processing. Establishing a new zero-carbon fuel production, supply, and marketing system in rural areas, converting these wastes into commercial energy, can potentially yield over more than a 50 percent increase in yield per mu for farmers.

High-quality new energy projects can effectively leverage investment opportunities, bring huge acceleration to China's new economy, and drive a large number of new economic growth points. The development of the new energy system in rural areas has given a boost to the related industries including PV, energy storage, power electronic devices, DC power distribution and protective devices, electric vehicles, and electric agricultural implements. Notably, trillion-level markets have emerged, encompassing rural new power system development, rural commercial fuel production-supply-marketing system development, rural smart energy operation system development, and rural electromechanical equipment standard battery development. That is to say, we take the road of socialist rural revitalization with Chinese characteristics, achieve industry revitalization through the energy revolution, implement the rural vitalization strategy in all aspects, complement agriculture with industry and propel rural development with urban development, and make agriculture and industry as well as urban areas and rural areas benefit and complement each other and develop and prosper together, thus speeding up the modernization of agriculture and rural areas.

2. Rural Eco-system Revitalization Backed by Energy Revolution

Eco-system revitalization is an intrinsic need of rural revitalization. The report of the 20th National Congress of CPC reaffirmed that green development is pivotal, aiming to maximize the value of the rural natural resources, and preserve pristine landscapes. After full electrification of the rural energy system, combustion of bulk coal and messy biomass is no longer seen, and sooty rooms and smoky air are the things of the past. The villages do not need to stockpile fuel or ash, which makes them look visually more pleasant. The straws and the livestock excrement are recycled and converted into commercial energy resources such as biogas or biomass pellets, and the conversion process also generates high-quality biomass fertilizers such as biogas residue and biomass ash. These fertilizers can be used in circular agriculture. This effort will eliminate the combustion of straws and further reduce fertilizer usage, thus contributing to the development of modern agricultural industrial parks and agricultural modernization demonstration zones.

As the first pilot county for the national rural energy revolution, Lankao County of Henan Province has achieved 3,170 h of clean energy-based power supply duration. The electricity generated by local renewable energy accounts for 87.48% of the electricity consumption of the whole county and accounts for 97.5% of the total energy production output of the county. Lankao County has comprehensively promoted cyclic utilization of domestic waste, crop straws, and livestock excrement, established the waste "collection-storage-transportation" system, and implemented systematic waste classification, leading to the effective consumption of various organic wastes in rural areas. The waste treatment capacity of the county is 600 t/d, and the excrement from large- and medium-sized livestock farms is effectively processed and converted, projects such as biomass-fueled thermal power plant, biogas plant, and organic fertilizer plant have been developed by using the local resources, the industry chain constituted of "planting-forage-breeding-gas production-biogas manure-planting" has been created, and the circular economy model defined by combined production of "gas, heat, electricity and fertilizers" has taken form. The innocent treatment rate of domestic waste has increased from 65 to 100%, the utilization rate of straws and livestock excrement exceeds 90%, and the ecological environment in the county territory has improved noticeably.

3. Rural Cultural Revitalization Backed by Energy Revolution

Cultural revitalization is a key cornerstone of rural revitalization and a key source of intrinsic driving power for rural development. The "culture" here refers to both energy supply and consumer culture. Most of the technologies and equipment required for the rural new energy system have matured, while some of them are still in the process of commercialization or the process of cost reduction and quality enhancement. The problems are not only from financial and institutional aspects, but also from energy supply and consumer culture reshaping.

Full electrification of rural areas and commercialization of biomass energy requires a cultural foundation, which encompasses the basic PV concepts, common knowledge on electricity use, safety-related knowledge, and conservation of biomass energy. Such knowledge must be instilled into the people from their childhood and incorporated into the curriculums of the primary and high schools, so as to popularize the basic culture and train the skilled personnel. Such knowledge must be extensively communicated to the rural dwellers, and some of them can start trying to use the new systems, so that the new energy supply and consumption culture will be eventually accepted by all of them. Cultural development may be the deciding factor that dictates the full implementation of the rural new energy system, and it is also a long-term undertaking that requires considerable efforts.

4. Rural Workforce Revitalization Backed by Energy Revolution

Workforce revitalization is the key to rural revitalization. The aging of the rural population, shortage of workforce, and failure to retain workforce are major constraints that hamper the development of agriculture, rural areas, and farmers. Through the energy revolution, the rural areas will improve their smart energy infrastructure, and create new power systems and zero-carbon fuel production, supply, and marketing systems. First, to maintain an efficient and stable rural energy system, a generation of local agricultural experts and professional agriculture managers must be trained. Second, the rural energy system will significantly increase revenues and mitigate the loss of the workforce. Third, with their abundant electricity and fuel and low land cost, the rural areas will incubate or attract a large number of energy-intensive enterprises such as fuel processing plants and agricultural products processing plants. This will create many local employment opportunities, and generate a number of rural energy experts who are familiar with rural development, energy technology, and rural management models. Therefore, the energy revolution will afford an adequate workforce to support the development of agriculture and rural areas.

5. Rural Organization Revitalization Backed by Energy Revolution

Organization revitalization is the fundamental support for rural revitalization. The primary CPC organizations must fulfill their role as the core leadership, promote party building to facilitate rural revitalization, improve the rural governance system that integrates autonomy, rule of law, and rule of virtue, and consolidate the governing base of the CPC in the rural areas. In 2019, Document No.1 of the Central Government stipulated that the village Party branch secretary must also perform the duties of the village committee director. Only by amassing wealth, can the villagers have a better future under the leadership of the primary CPC Organization. In reality, many villages lack a source of revenue and can only rely on the fiscal appropriation of the higher-level government, so the villagers cannot take a deciding role or leading role in village governance.

The revenue generated by the electricity sale of village poverty alleviation PV power stations will strengthen the village collective economy and increase the annual dividend of the villagers, and such power stations will provide the villagers with employment positions in cleaning and maintenance. Since 2014, the country has enacted many PV-based poverty alleviation policies, and has built poverty alleviation PV power stations with a total capacity of 26,360,000 kW that benefit 60,000 poverty-stricken villages and 4,150,000 poverty-stricken households, generate about RMB 18 billion revenue and provide 1,250,000 employment positions. In 2019, a DG roof PV power generation project with a total investment of RMB 644,000 and a total capacity of 117 kW was built at three villages that had proper conditions, namely Dangcheng Village, Chengguan Village and Chengbei Village of Dangchengwan Town, Subei County, Jiuquan, Gansu. Each of these villages gained an annual income of RMB 20,000–30,000, and they changed from "empty villages" to "wealthy villages". This is a PV-based poverty alleviation model that achieves rural revitalization through the green new energy industry and generates both economic and environmental benefits. More importantly, the villagers gain stable income and the primary CPC organizations consolidate their cohesiveness.

4.7 Summary

The primary objective of China's energy revolution is to establish a zero-carbon power system and a zero-carbon fuel production, supply, and marketing infrastructure. Given the abundant spatial and biomass resources in rural areas, they will transition from being energy consumers to becoming energy suppliers in the future zero-carbon energy framework. The rural areas will lead in achieving carbon neutrality and play a pivotal role in shaping the zero-carbon energy system, thus making significant contributions. Consequently, the development of a rural new energy system becomes an integral component of China's overall energy transitions.

The rural energy system is mainly designed as a new PEDF distributed system underlain by a rooftop PV system, and is also meant to increase the proportion of commercialized biomass energy products. As estimated, the current rooftop space in rural areas is enough to install about 2 billion kW PV capacity and generate 2.9 PWh every year. Therefore, fully exploiting the rural rooftop space resource will provide the energy required for living, production, and transportation, and the electricity generated will supersede coal, oil, gas, and biomass, thus realizing a complete zero-carbon economy in rural areas. The rural areas will be the most important base of zero-carbon fuels, and the biomass energy from them will account for over 70% of the future energy demand. Under ideal conditions, the rural areas will output 2.3 PWh of zero-carbon electricity and 740 million tce of zero-carbon fuel in 2060.

The rural areas will be the first areas to establish the new energy system and thus make a great attempt in the national energy transition. Through continuous practice and exploration, we will test various technologies, policies, and mechanisms of the distributed energy system, identifying feasible practices to implement in rural and urban areas alike, facilitating an energy transition from households to cities.

The rural new energy transition is estimated to require a total economic investment of RMB 10 trillion. It is recommended that the relevant departments prioritize the development of a new rural energy system to solve the issues relating to agriculture, rural areas, and farmers, and accomplish rural revitalization, and conduct top-level design and planning in this regard. The "Village Wind Power Program" and "Household PV Program" must be implemented steadily, the county-level rooftop distributed PV system projects must proceed in an orderly fashion, the development of renewable energy must be intensively integrated with ecological civilization development, new urbanization model, new infrastructure and new technologies, new fields and new scenarios of renewable energy development must be explored, and development of the zero-carbon villages, towns, and counties will be established phase by phase.

Open Access This chapter is licensed under the terms of the Creative Commons Attribution-NonCommercial-NoDerivatives 4.0 International License (http://creativecommons.org/licenses/by-nc-nd/4.0/), which permits any noncommercial use, sharing, distribution and reproduction in any medium or format, as long as you give appropriate credit to the original author(s) and the source, provide a link to the Creative Commons license and indicate if you modified the licensed material. You do not have permission under this license to share adapted material derived from this chapter or parts of it.

The images or other third party material in this chapter are included in the chapter's Creative Commons license, unless indicated otherwise in a credit line to the material. If material is not included in the chapter's Creative Commons license and your intended use is not permitted by statutory regulation or exceeds the permitted use, you will need to obtain permission directly from the copyright holder.

Chapter 5
Potential and Characteristics of Rural Distributed Zero-Carbon Energy in China

5.1 Definition of Rural Distributed Zero-Carbon Energy

"Zero-carbon" means that the energy production and consumption processes will not increase carbon dioxide emissions. Zero-carbon energy can come from a wide range of sources, including solar energy, wind energy, nuclear energy, hydraulic energy, and biomass energy, etc. However, the aforementioned energy sources are not all commonly available in rural areas, so rural zero-carbon energy must be developed based on geographic characteristics.

The rural energy sources are characterized by scattered and extensive distribution, while the large-scale PV power stations, hydropower stations, wind power plants, and nuclear power plants all use concentrated production capacity and are directly connected to the power grid. Though certain stations or plants are located in remote locations, they are not connected to the rural energy network, these energy sources are not counted as rural zero-carbon energy. There is also the zero-carbon energy represented by geothermal energy. This energy is extensively found in rural areas, but exploitation of this energy and maintenance of the facilities are quite difficult, so this energy is infeasible for rural areas. As a result, rural areas should prioritize development of the zero-carbon energy sources that are easy to access and practicably meet the energy demand. The most typical energy sources are rural rooftop PV power, biomass energy, and small hydropower stations, as shown in Fig. 5.1.

Rural rooftop PV power, biomass energy, and small hydropower stations are highly abundant zero-carbon energy sources that have not been adequately exploited, and they possess vast potential for future development under the carbon neutrality goal. Therefore, this chapter proposes the methods for assessing the potential of the aforesaid energy sources, and analyzes the total quantity and spatial distribution of these energy sources across the country.

Fig. 5.1 Typical distributed zero-carbon energy sources in rural areas. **a** Rural rooftop PV; **b** biomass; **c** rural small hydropower stations

5.2 Analysis of Total Quantity and Distribution of Rural Rooftop PV Resources in China

5.2.1 Important Value of the Rural Rooftop PV Resources

Solar energy, characterized by wide distribution, cleanness, and sustainability, has always been an important one of many various renewable energy sources. In 2019, the global solar energy consumption was 4.1 EJ [9], which accounted for 1.1% of the global total energy consumption, but the solar radiation received by the planet Earth every year is 4 million EJ [10], so this energy source still has great potential for exploitation. With the cost of solar cells significantly reduced [11], PV power generation is the most important way of using solar energy. The global total installed capacity of PV power generation increases by nearly 40% every year [12], and reached 760 GW in 2020 [13]. At present, the total installed capacity of PV power in China is 205.7 GW. By 2030, the total installed capacity of wind power and solar power will exceed 1.2 billion kW [14]. In recent years, the rooftop PV system has drawn more and more attention, because PV systems can be installed in the locally available rooftop space to meet the energy demand of the households, and to reduce the power loss in the transmission and conversion processes. Exploitation of the rooftop space will reduce the land space occupied by PV systems, and the PV panels can be installed as building materials for new buildings or as part of the modification of existing buildings [15, 16].

PV systems installed in rural rooftop space have their unique advantages. The roofs in the rural areas are less shadowed and they are regularly shaped, which makes them conducive to installing PV panels. The rural areas have low population density and single-household buildings take up a high proportion of all the buildings, so the power generated by rooftop PV system is enough to meet the power demand of the households. After firstly self-use in the households, and the surplus power can be diverted to the power grid. This will eliminate the traditional inefficient and pollutive energy sources such as bulk coal and biomass and significantly improve the living quality of the rural households in developing countries [17]. The DC microgrid system underlain by a rooftop PV system can contain the energy generated by the

5.2 Analysis of Total Quantity and Distribution of Rural Rooftop PV ...

rooftop power generation system, and the inter-household DC bus will increase the effective utilization of PV power and minimize power loss. The village microgrid can also be connected to the municipal power grid [18]. The PV system can not only meet the power demand of households, but also provide renewable electricity to irrigation pumps and small industrial entities. Due to the low levels of energy consumption in rural areas, the surplus electrical power can be diverted to the urban areas. In light of this, rooftop PV will be an important development direction of rural renewable energy in China.

5.2.2 Method for Analyzing the Power Generation Potential of Microscopic Rooftop PV System

1. Rural Roof Types Identified by Using Remote Sensing Satellite Image

The fundamental issue of rural rooftop PV resources is the utilization of the rooftop space. Unlike urban buildings, rural buildings are quite simple in form, and most roofs are flat roofs or pitched roofs. The rural areas have other roof types such as single-slope roofs or four-slope roofs, but such roofs take up only a low proportion, so they can be essentially neglected. Any roof with a gradient less than $10°$ is defined as a flat roof, while the other ones are defined as pitched roofs. The purpose of identifying rural roof types is to calculate their PV power generation potential, so the roofs are classified with consideration of the PV panel installation methods for flat roofs and pitched roofs. For a flat roof, all the roof area can be used for installing PV panels, minus the space needed to offset mutual shadowing between contiguous PV panels and the reserved spacing for maintenance. For a pitched roof, the angle and direction of the roof must be taken into consideration, where a roof with a $<45°$ included angle between its direction and the due south direction is defined as a southward pitched roof, while any roof with such included angle of $>45°$ is defined as an east–west pitched roof. For a southward-pitched roof, only the surface exposed to the sun can be used for installing PV panels. For an east–west pitched roof, all the roof surfaces can be used for installing PV panels, but the east and west sides receive different levels of radiation. Figure 5.2 shows a typical rural building and the aforesaid roof types. By now, three roof types have been determined to be identified through remote sensing satellite images, including flat roof, southward pitched roof, and east–west pitched roof.

2. Source of Remote Sensing Satellite Image Data

The existing commercial satellites can generate panchromatic images with 0.15 m precision, but the high cost of such satellite images limits their extensive application, so two types of open-source satellite images have been used. Figure 5.3 shows the same village presented in two types of images. The two images are quite close temporally and the rural buildings change less, so the two images can be both used for identifying the roofs. The above images are both resampled to 0.6 m resolution

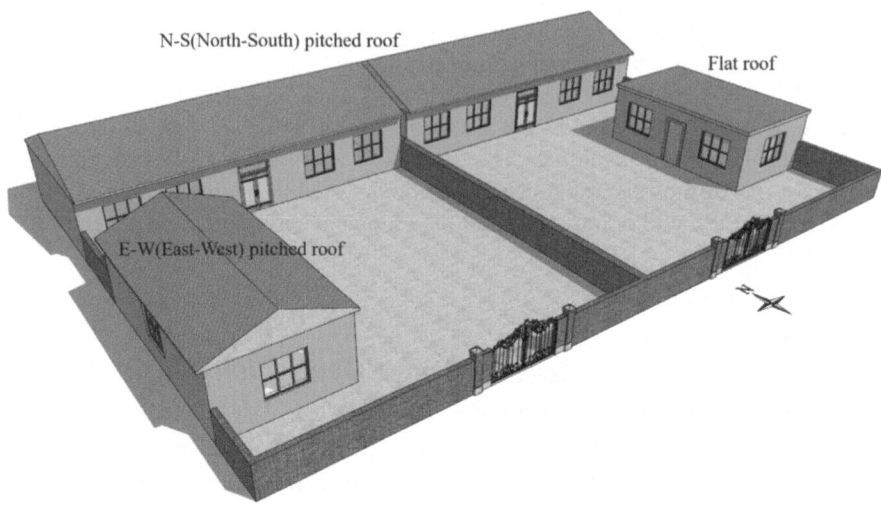

Fig. 5.2 Schematic diagram for typical rural building

Fig. 5.3 Two types of open-source remote sensing satellite images and the actual labels for the corresponding roofs

by using the geographic information system (GIS), and then they are superposed to enhance the pixel features of the roofs, thus making them easier to identify later.

The labels for the three roof types are also shown in Fig. 5.3. The green label represents a southward pitched roof, the blue label flat roof, the yellow label east–west pitched roof, and the black label is the background value. The objects to be identified are the three types of rural roofs, so the other objects such as cropland, water bodies, and forest land are labeled as background values.

3. Extraction of Rural Roofs

The process of extracting the pixels representing the various rural roofs from the remote sensing satellite images is a study subject of semantic segmentation. One of the most representative studies in the area of semantic segmentation is the U-Net model created by Ronneberger [19] et al. in 2015. This model was initially used for

5.2 Analysis of Total Quantity and Distribution of Rural Rooftop PV …

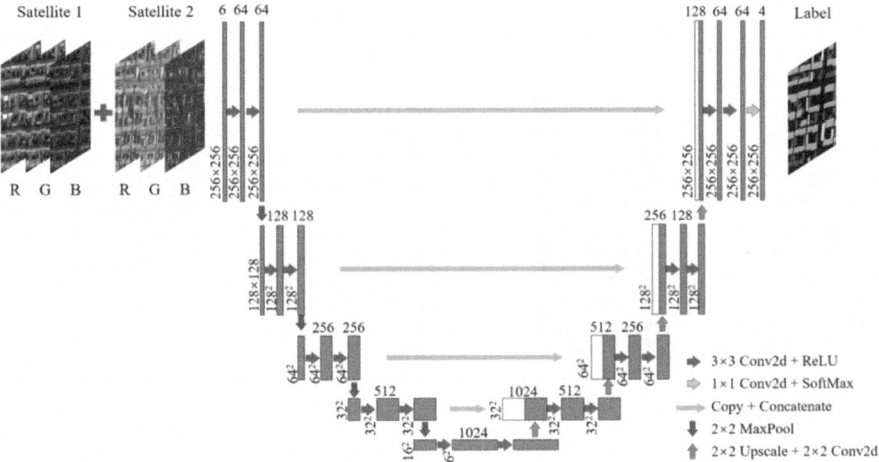

Fig. 5.4 Structural diagram for corrected U-Net model

cellular image segmentation in the area of biomedical imaging. The U-Net model has a simple and clear structure and excellent performance in image segmentation, and it requires less training samples and time than the other semantic segmentation models. Therefore, this study used a classical U-Net network as the basic model. By adjusting the model parameters, we detected different types of rural rooftops from the satellite images.

Figure 5.4 shows the corrected model. To enhance the image features of the rural roofs, two types of remote sensing satellite images are superposed to generate the 6-channel satellite images that are used as the model input. All the convolutional functions used the zero-padding method to ensure that the output images are of the same size as the input images. The output image contains four types of pixels, including flat roof, southward pitched roof, east–west pitched roof, and background.

To train the aforesaid model, the maximum likelihood equation is used to calculate the probability of the classification of each pixel in the input image. The cross-entropy loss function is used as the objective function, and the gradient descent algorithm is used to train the various layers of parameters of the model. After model training is completed, the model can be used to extract the various types of rural roofs from the aforesaid open-source satellite images.

4. Determining the Power Generation Potential of Rural Rooftop PV

The identification result obtained by using the corrected U-Net model is the classification of each pixel in the image. However, the purpose of identification is to obtain the size of the roof that is used to calculate the installation area for PV panels, so the identification result is vectorized first, as shown in Fig. 5.5. The perimeter and area of the identified roof can be used to calculate the equivalent length and width of the roof, and the identification result is regularized to make it more approximate to the actual value.

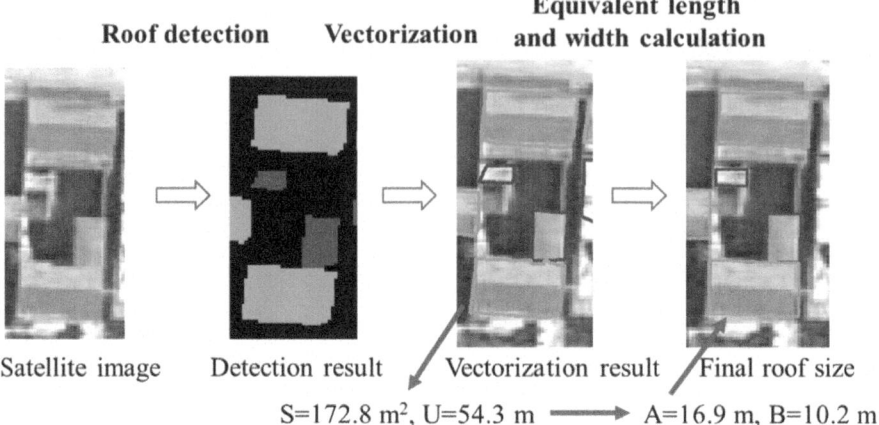

Fig. 5.5 Processing of identifying roofs. U-roof perimeter; S-roof area; A-equivalent length of roof; B-equivalent width of roof

After the sizes of the various roofs are determined, the area available for installing PV panels can then be calculated. Two methods are usually used for installing PV panels on roofs, i.e., optimum tilt angle installation and parallel roof installation. Taking a rural flat roof as an example, if the optimum tilt angle installation method is used, the PV panel is installed on the roof with the tilt angle that enables the maximum annual generated power, and the mutual shadowing between the contiguous rows of PV panels is also taken into account, as shown in Fig. 5.6a. If the parallel roof installation method is used, the mutual shadowing and the spacing between PV panels are not considered, as shown in Fig. 5.6b. The optimum tilt angle installation method allows the energy production per unit area of the PV panel to be maximized, thus coming with a better cost–benefit ratio. The parallel roof installation method allows the PV panel installation area to be maximized, thus coming with greater power generation potential.

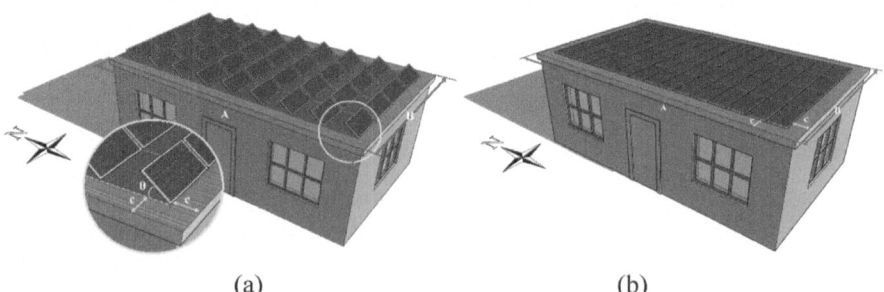

Fig. 5.6 Schematic diagram for two installation methods for rooftop PV panel on a flat roof. **a** Optimum tilt angle; **b** parallel roof installation

5.2 Analysis of Total Quantity and Distribution of Rural Rooftop PV …

For the southward-pitched roof and east–west-pitched roof, the PV panel installation area can be determined similarly. For a southward-pitched roof, only the availability of the surface exposed to the sun is considered. For an east–west pitched roof, the shadowing caused by both the PV panel and the roof tilt angle needs to be considered.

After the PV panel installation area of each type of roof is calculated, the power generation potential of rooftop PV can be calculated by determining the solar radiation received by the PV panel, the PV module efficiency and PV performance ratio.

5. Result of Analysis of Typical Rural Rooftop PV

The aforesaid corrected U-net neural network model is applied to the satellite image of a town in North China, and a field investigation is conducted at the 10 villages in that area, to obtain the actual labels of the buildings. The aforesaid information is used to generate a dataset that is used to train the U-Net model. Two representative rural households are selected in the field investigation and the roof area is measured, and the trained model is applied to the satellite image of the two rural households to identify the types of roofs, as shown in Fig. 5.7. According to the comparison, the roof area identified by the model proposed herein is quite approximate to the measured value, and the roof types are classified correctly.

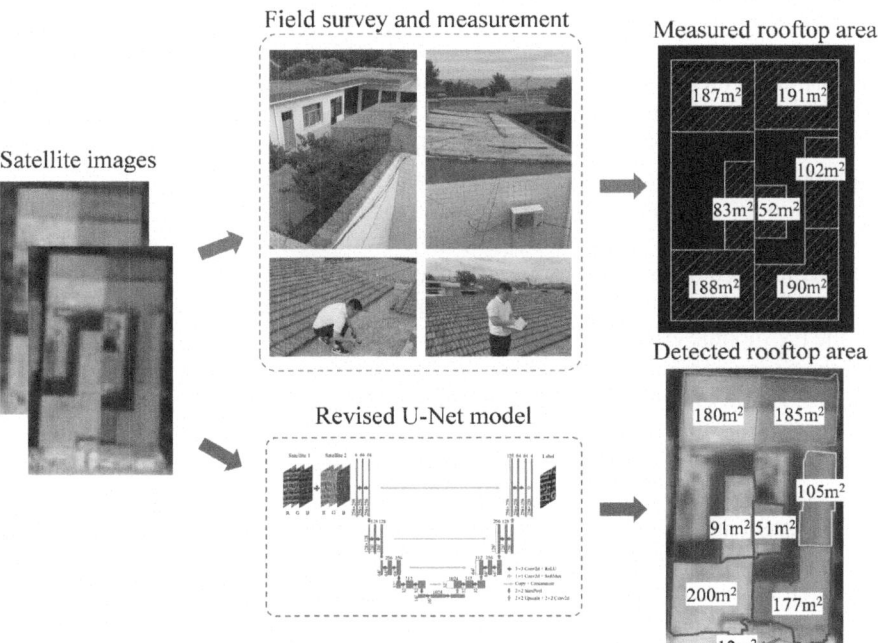

Fig. 5.7 Comparison between identification result and measurement result

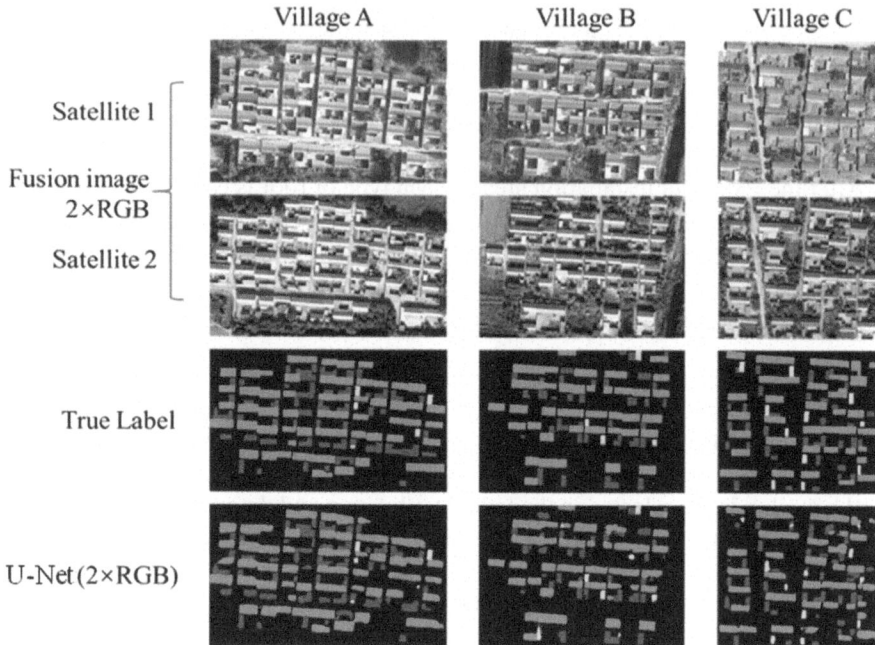

Fig. 5.8 Identification result of the corrected U-Net model

The trained model is applied to the satellite images of the three surveyed villages, to obtain the roof identification result shown in Fig. 5.8. Two types of satellite images and roof labels are generated, to identify the objects in the output images. As the result proves, the model proposed herein can extract various rural roof areas from the satellite images and thus make it feasible to analyze the rooftop PV potential.

For the various types of roofs identified, the available area for installing PV panels can be calculated by using the aforementioned methods. The solar radiation received by the PV panel installed in different manners is queried from the software database, and the PV module efficiency η is set as 16% and the PV performance ratio λ as 85%, then the power generation potential of the PV panels on each roof can be calculated. Figure 5.9 shows the power generation potential of the rooftop PV system in village A, and specifically shows the potential achieved by each of the two installation methods. The potential distribution diagram shows the space coordinates and scale. This coordinate is the global geographic coordinate in the WGS1984 projected coordinate system and is given in m, so the PV power generation potential of each roof accurately matches the spatial position. The parallel roof installation method allows more area for installing PV panels, so it achieves greater power generation potential even though the solar radiation received is less than the optimum tilt angle installation method. The power generation potential of the individual buildings in village A is quantified, the result is that the maximum value exceeds 40,000 kWh and the average value is 16,900 kWh.

5.2 Analysis of Total Quantity and Distribution of Rural Rooftop PV …

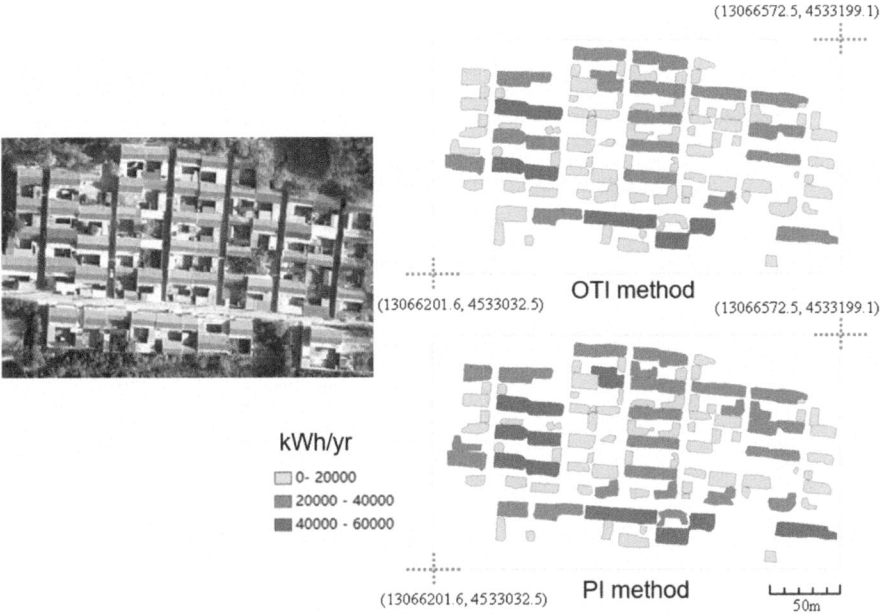

Fig. 5.9 Calculated rooftop PV power generation potential of village A

The rooftop PV power generation potential of a whole village is the accumulation of those of all the households. This method is then applied to all the villages investigated, to calculate their total PV power generation potential, and then the total value is divided by the number of households in these villages to determine the average rooftop PV power generation potential of the individual households.

Table 5.1 indicates the PV power generation potential of the individual households in the study area. If the parallel roof installation method is used, the annual total PV power generation potential of the households is 26,700–46,300 kWh, 36,200 kWh on average. If the optimum tilt angle installation method is used, the total value is 20,000–33,600 kWh, 26,500 kWh on average. In fact, however, the annual power consumption of a rural household in a well-developed area is still less than 5,000 kWh [20, 21], so the annual total energy production of a single household far exceeds its annual power consumption, no matter which installation method is used.

The rooftop PV power generation potential of individual households calculated by the aforesaid method can be accumulated in various ranges. The spatial distribution of the households in these villages is given above. If the study scope is expanded to a town, as shown in Fig. 5.10, the rooftop PV power generation potential of the villages in the town is calculated. This figure illustrates the space coordinates and scale. The coordinate is the global geographic coordinate in the WGS1984 projected coordinate system and is given in km. To avoid confusion, this figure only shows the power generation potential achieved by the parallel roof installation method for PV panels. The size of the circle represents the potential of the village. As can be seen,

Table 5.1 Annual total PV power generation potential of the villages and households in the study area

Village	Number of households	Annual total power generation potential of the whole village (10,000 kWh)		Annual total power generation potential of a household (10,000 kWh/household)	
		Parallel roof	Optimum tilt angle	Parallel roof	Optimum tilt angle
A	63	179	136	2.84	2.16
B	39	104	78	2.67	2.00
C	36	129	96	3.59	2.68
D	879	2927	2110	3.33	2.40
E	240	974	696	4.06	2.90
F	215	808	600	3.76	2.79
G	470	1725	1288	3.67	2.74
H	188	694	523	3.69	2.78
I	101	468	339	4.63	3.36
J	62	226	165	3.64	2.66
K	79	359	257	4.55	3.25
Average				3.62	2.65

the PV power generation potential of the villages in the study area varies between 1.7 and 29.3 million kWh, approximately 10 million kWh. With the optimum tilt angle installation method, the average power generation potential is 7.3 million kWh. The aforementioned power generation potential and the architecture of the existing power grid can be used together for renewable power system planning and other scenarios.

5.2.3 Analysis of the Power Generation Potential of Macroscopic Rooftop PV System

For macroscopic rooftop PV systems or even the rooftop PV resources of the entire country, the key is also to extract the rooftop area. If the information is to be extracted by using high-resolution satellite images, the excessively large amount of data processed and the difficulty in applying computer vision models to different regions will make large-scale identification impossible. The low-resolution satellite image can support large-scale identification, but it is difficult to distinguish the roofs and the identification accuracy is too low. With the comprehensive assessment of the characteristics of the satellite images of different resolutions, a nationwide rural rooftop PV potential calculation method based on the proportion coefficient is proposed here, as shown in Fig. 5.11. The calculation process of this method is explained below.

5.2 Analysis of Total Quantity and Distribution of Rural Rooftop PV …

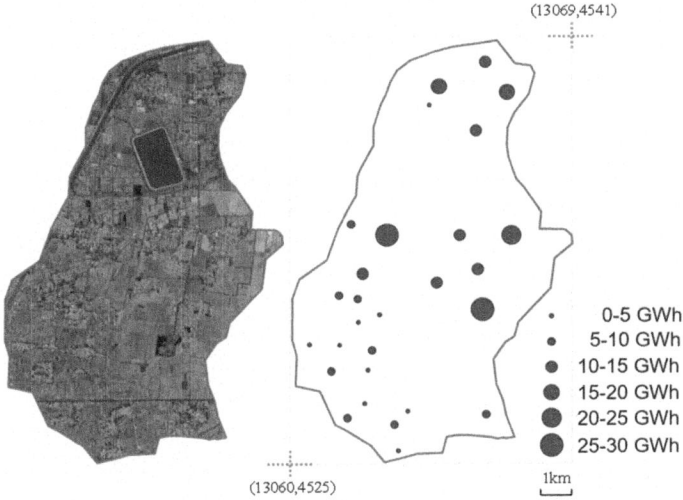

Fig. 5.10 Calculated rooftop PV potential of the villages in the town territory

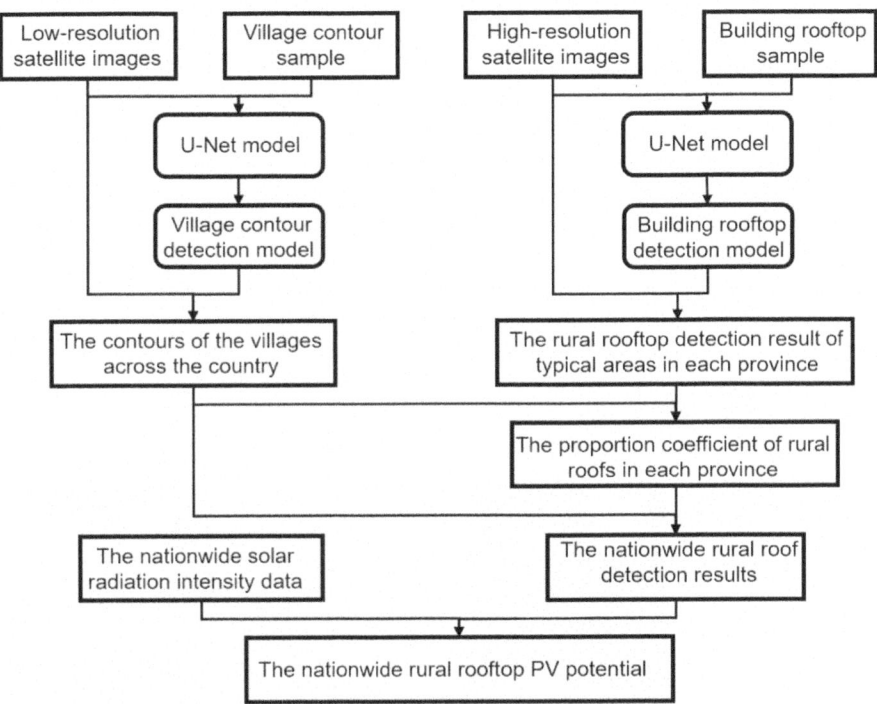

Fig. 5.11 Study process for nationwide rural rooftop PV resource potential

(a) The contour of the village is extracted by using a low-resolution satellite image. It is difficult to identify roofs in a low-resolution satellite image, so the contour of the entire village is identified instead. A dataset is created for each area and then the model is trained and applied, so as to identify the contours of the villages across the country.
(b) Typical areas are sampled from the provinces and municipalities across the country, the roofs in all the villages in these areas are identified by using high-resolution satellite images, and then the roof identification result is combined with the contour identification result, to calculate the proportion coefficient of the rural roofs in each province.
(c) The proportion coefficient of the rural roofs is then multiplied by the nationwide village contour identification result to obtain the nationwide rural roof area.
(d) Then, based on the proportion of each type of roof in different regions and the local solar radiation intensity, the potential installed capacity and energy production of the nationwide rural rooftop PV systems can be calculated.

1. Village Contour Extracted by Using Low-resolution Satellite Image

The village contour is the external contour of the residential area, and it encompasses the RRB roofs, courtyards, and roads. The process of identifying village contour is also done by using the computer vision method, where only the villages and other elements need to be identified, so the classic binary U-net neural network is used. Samples are collected from different regions in the country, and the village scope is plotted manually and used as a training label. Figure 5.12 shows typical northern and southern villages and their label samples. The northern villages are relatively isolated and they are in regular shapes and directions. The southern villages are mostly built along roads and are in irregular shapes. The aforementioned labeled images experience input size cropping and data augmentation to generate the training set and test set. Then sample data training is done by using the U-net convolutional neural network. It must be noted that different models are trained to identify the village contours because the village form and image vary greatly across different regions in the country.

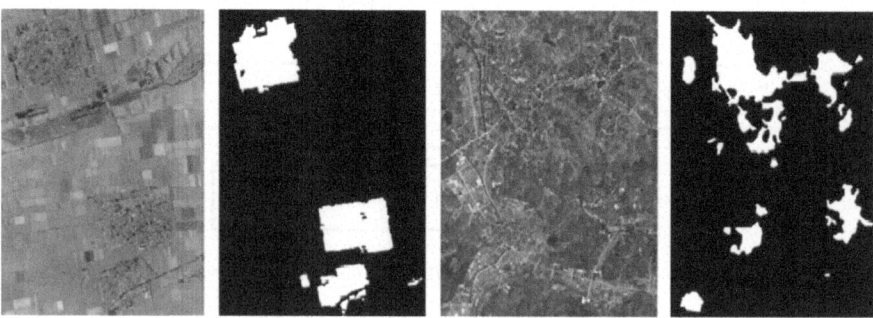

Fig. 5.12 Satellite image and country label of typical villages

5.2 Analysis of Total Quantity and Distribution of Rural Rooftop PV ...

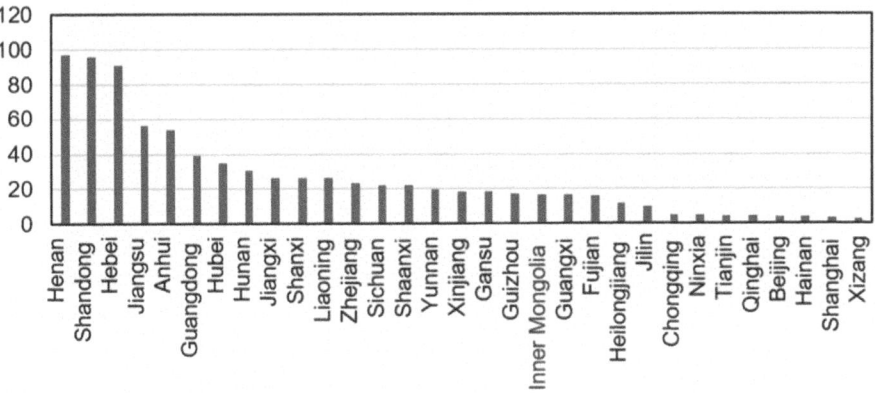

Fig. 5.13 Comparison of different provinces in terms of village contour area

Contour identification of all the villages in the country is completed by using satellite images of 2 m/pixel resolution, and the total identified area of the provinces is about 82 billion m². As can be seen in Fig. 5.13, the densely populated regions such as Henan, Shandong, and Hebei have large village contour area, which accounts for about 34.5% of the total area of these regions. While the less populated regions such as Qinghai and Xizang and the municipalities of small areas have fewer villages and thus have less contour area. However, the village contour area cannot be directly used to calculate the rooftop PV potential, but has to be converted into the roof area first.

2. Determining the Proportion Coefficient of Rural Roofs in Different Provinces and Municipalities

The proportion coefficient of rural roofs is defined as the ratio R_c between the roof area of the households and the total contour area of the village, as calculated by Eq. (5.1), where S_p is the projected area of household roof, S_D is the village contour area, and R_c is the ratio. The household roofs can be identified from high-solution satellite images. Figure 5.14 shows the contour identification result and rural roof identification result of typical northern and southern villages. To increase the calculation efficiency, rural roof identification is also done by using the conventional binary U-Net model.

$$R_c = S_p/S_D \qquad (5.1)$$

Figure 5.14 shows two typical villages. To make the proportion coefficient of the rural roofs universally applicable, the proportion coefficient of multiple villages in one region needs to be calculated and then their average is also calculated; furthermore, the physical difference between the villages has to be taken into account, and the proportion coefficient of the rural roofs in different regions is also calculated. The following principle is followed: certain villages are selected from the list of traditional

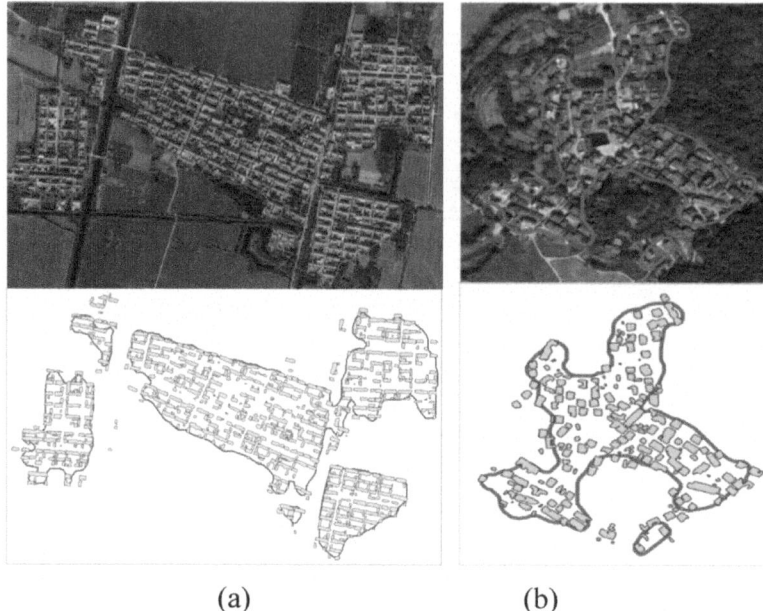

(a)　　　　　　　　　　(b)

Fig. 5.14 Contour identification result of typical villages and rural roof identification result. **a** Northern village; **b** southern village $R_c = S_p/S_D (4\text{--}1)$

villages identified, and the mountainous villages, coastal villages, flatland villages, highland villages, suburban villages, and forest villages are also identified. This process covers all types of rural residences. Besides, all the villages selected are the ones with independent organizational systems and clear boundaries, so that the rural roof samples selected are representative and the selection and calculation processes can be made easier. For the sample villages selected, the roofs are identified from high-resolution images, and then the identification result is combined with the result of nationwide village contour identification to calculate the proportion coefficient of the rural roofs in each region.

Then the proportion coefficient of the buildings in all the sampling regions is calculated by using the aforesaid method, and the results are listed in Table 5.2. The proportion coefficient of the rural roofs in the provinces is 0.19–0.44. This coefficient is high in densely populated regions, and it is low in sparsely populated regions such as Gansu and Xinjiang. A total area of 22,436,000 m² of roofs has been identified in the sampling villages, and the contour area of these villages is 73,220,000 m². The average proportion coefficient of the rural roofs in China is approximately 0.31, i.e., about 31% of the contour area identified from the low-resolution images is roof area.

The proportion coefficient of the rural roofs of the sampling regions is multiplied by the rural village contour area identified, to obtain the identified rural roof area of the regions, also as shown in Table 5.2. The calculated total rural roof area of the regions is about 27.33 billion m². The provinces with the greatest areas are Henan and

5.2 Analysis of Total Quantity and Distribution of Rural Rooftop PV …

Table 5.2 Calculated proportion coefficient of rural buildings and roof area in some of the provinces (autonomous regions or municipalities) in China

Province (autonomous region and municipality directly under the Central Government)	Sample regions			All regions	
	Village contour area (10,000 m^2)	Roof area (10,000 m^2)	Proportion coefficient of buildings	Village contour area (100 million m^2)	Roof area (100 million m^2)
Beijing municipality	72.7	29.2	0.40	3.9	1.6
Tianjin	273.0	88.4	0.32	4.4	1.4
Hebei province	176.1	54.5	0.31	90.6	28.0
Shanxi province	1287.7	524.1	0.41	26.1	10.1
Inner mongolia autonomous region	72.4	18.2	0.25	16.6	4.2
Liaoning province	396.8	129.3	0.33	26.0	8.5
Jilin province	159.6	46.0	0.29	10.0	2.9
Heilongjiang province	36.7	15.0	0.41	11.5	3.9
Shanghai municipality	6.8	2.4	0.36	2.9	1.0
Jiangsu province	59.2	20.2	0.34	56.0	19.1
Zhejiang province	382.8	129.4	0.34	23.5	7.9
Anhui province	71.2	14.9	0.21	53.9	11.3
Fujian province	699.1	187.3	0.27	16.1	4.3
Jiangxi province	54.3	19.5	0.36	26.6	9.5
Shandong province	657.1	227.9	0.35	95.6	33.2
Henan province	52.3	19.6	0.38	97.0	36.3
Hubei province	140.3	35.7	0.26	34.6	9.0
Hunan province	205.9	74.2	0.36	30.6	11.0
Guangdong province	196.6	86.9	0.44	39.3	17.4
Guangxi Zhuang Autonomous region	27.8	15.8	0.39	16.4	6.5

(continued)

Table 5.2 (continued)

Province (autonomous region and municipality directly under the Central Government)	Sample regions			All regions	
	Village contour area (10,000 m²)	Roof area (10,000 m²)	Proportion coefficient of buildings	Village contour area (100 million m²)	Roof area (100 million m²)
Hainan province	40.2	14.3	0.36	3.9	1.4
Chongqing municipality	11.9	5.8	0.41	5.3	2.2
Sichuan province	111.6	54.2	0.41	21.9	8.9
Guizhou province	92.8	35.5	0.35	17.4	6.1
Yunnan province	50.2	19.4	0.39	19.5	7.5
Xizang autonomous region	161.8	32.5	0.20	2.5	0.5
Shaanxi province	235.5	93.7	0.40	21.8	8.7
Gansu province	392.3	75.3	0.19	18.1	3.5
Qinghai province	223.6	90.9	0.41	4.3	1.7
Ningxia Hui Autonomous region	46.7	15.3	0.33	5.2	1.7
Xinjiang Uygur Autonomous region	1051.6	140.7	0.21	18.7	3.9
Total	7322.0	2243.6	0.31	820.2	273.3

Shandong, which match the rank of the rural population in the statistics yearbook. Guangdong, which has the third-largest population, has a high proportion of rural multi-story buildings, so its roof area is less than that of Hebei which has the fourth-largest population. In short, the aforesaid method can be used to directly calculate the rural roof area in the regions and determine the spatial distribution of the villages.

3. Calculating the Rural Rooftop PV Power Generation Potential of the Regions

Based on the proportion of each type of building in the sampled regions, the rural rooftop PV power generation potential of the provinces in the country can be calculated, as shown in Table 5.3. The available rural roof area for installing PV panels in the regions is 13.12 billion m² in total, the potential installed capacity is 2 billion kW,

5.2 Analysis of Total Quantity and Distribution of Rural Rooftop PV ...

Table 5.3 Potential rooftop PV installation area, energy production, and installed power in the rural areas in some of the provinces (autonomous regions or municipalities) in China

Province (autonomous region and municipality directly under the Central Government)	Available rural roof area for installing PV panels (100 million m^2)	Annual energy production (100 million kWh)	Installed capacity (10,000 kW)
Beijing municipality	0.8	175.6	1178.7
Tianjin	0.7	157.1	1054.6
Hebei province	13.5	3091.9	20,759.2
Shanxi province	4.9	1190.1	7992.6
Inner Mongolia autonomous region	2.0	541.1	3633.9
Liaoning province	4.1	976.0	6554.0
Jilin province	1.4	324.3	2177.4
Heilongjiang province	1.9	415.3	2788.6
Shanghai municipality	0.5	108.1	725.9
Jiangsu province	9.2	2074.2	13,928.8
Zhejiang province	3.8	816.0	5478.4
Anhui province	5.4	1142.1	7669.7
Fujian province	2.1	487.9	3276.6
Jiangxi province	4.6	1014.8	6814.9
Shandong province	15.9	3799.5	25,515.6
Henan province	17.4	3860.6	25,927.1
Hubei province	4.3	826.0	5546.6
Hunan province	5.3	1013.5	6804.5
Guangdong province	8.3	1859.7	12,488.3
Guangxi Zhuang autonomous region	3.1	633.0	4250.8
Hainan province	0.7	192.8	1294.4
Chongqing municipality	1.0	157.2	1055.9
Sichuan province	4.3	801.1	5379.5
Guizhou province	2.9	544.5	3655.3
Yunnan province	3.6	974.9	6547.5
Xizang autonomous region	0.2	74.8	506.6
Shaanxi province	4.2	915.9	6149.9
Gansu province	1.7	462.7	3107.4
Qinghai province	0.8	247.6	1662.9
Ningxia Hui autonomous region	0.8	222.3	1492.8

(continued)

Table 5.3 (continued)

Province (autonomous region and municipality directly under the Central Government)	Available rural roof area for installing PV panels (100 million m^2)	Annual energy production (100 million kWh)	Installed capacity (10,000 kW)
Xinjiang Uygur autonomous region	1.9	508.3	3413.3
Total	131.2	29,608.8	198,831.6

and the estimated annual energy production is 2.9 PWh. Given that the current coal consumption index in the electrical power industry is 301.5 gce/kWh, such energy production is equivalent to about 892 million tce.

The decisive factor influencing rural rooftop PV resources is the roof area. The northwest and northeast regions have higher solar radiation intensity than Central China but have less roof area, so they have less power generation potential. Henan, Shandong, and Hebei rank the first three places in terms of rural rooftop PV installed capacity and total energy production, and their total potential installed capacity is 720 million kW and annual power generation potential of 1.08 PWh, accounting for 36.4% of the national total.

According to the above analysis, the rural rooftop PV resources in China have immense power generation potential, comparing with the domestic power consumption. In 2018, for example, the domestic power consumption in the rural areas in China was 26,230,000 kWh, while the annual power generation potential of the rural rooftop PV resource was 2.9 PWh, which was more than 10 times the former. Therefore, rational exploitation of the rural rooftop PV resources will more than suffice to meet the power demand of the rural areas and also provide zero-carbon electricity to the urban areas.

5.3 Analysis of the Biomass Resources of the Rural Areas in China and Their Distribution

5.3.1 Important Value of the Rural Biomass Resources

Biomass is a form of energy stored in living organisms after solar energy is converted into chemical energy by photosynthesis. The solar energy stored through photosynthesis on the planet Earth every year is 10 times the global annual energy consumption. Such energy exists in various forms of biomass energy, and these biomass resources are abundant and diverse. The most important and most commonly found biomass energy exists in the agricultural residual materials such as straws and the residual materials from forestry. According to statistics, the amount of agricultural residual materials generated by the 7 main countries and regions (China, the United

States, EU, Brazil, India, Argentina, and Canada) is 3.3 billion tons now, and the forest biomass generated in the world every year is about 4.6 billion tons. The amount of usable residual materials is about 700 million tons [22]. Though biomass resources exist in high total abundance, the biomass energy consumed in the world in 2017 was about 790 million tce, accounting for only 9.5% of the primary energy consumed in that year, and more than half of that energy was used for traditional biomass-fueled kitchen work, heat supply, and small industries according to the statistics of IEA. Therefore, the current utilization efficiency of biomass energy is quite low and there are still many resources unexploited.

Compared with other renewable energy sources such as solar energy and wind energy, biomass energy has the following properties: ① Biomass energy is by far the only zero-carbon emission renewable energy that can be directly obtained and used as fuel, and this feature should be fully utilized. ② Biomass energy can be stored and transported at low cost and can be converted into adjustable and controllable energy through certain processes. ③ Biomass energy is a highly "influencing" energy, which is seen in that biomass fuel production can expand agricultural production, boost rural economic development, facilitate agricultural modernization of grain production areas, increase the income of farmers, help develop clean biofuel and biomass power generation, boost agricultural industrialization and the development of medium or small towns, and close the gaps between urban and rural areas. ④ Biomass is a zero-carbon emission energy if it is efficiently used as a fuel, but if it is discarded or returned to the farmland, it will generate large amounts of greenhouse gases such as CO_2, CH_4, and N_2O, which will aggravate carbon emission. Therefore, rational utilization of biomass is very important to China's efforts to cope with climate change and reduce carbon emissions.

The EU countries take much account of biomass energy development, and have promoted its development through legislation, tax reduction, fiscal subsidies, and other means in recent years. For example, Sweden has enacted tax exemption policies and allocates EUR 36 million of governmental budget every year to support the development of biomass fuels and conversion technologies; Germany grants low-interest loans to biomass energy development projects; the EU ratified the *European Green Deal* (EGD) in 2019 [23], and it will invest EUR 1 trillion in renewable energy development in the following 10 years. Biomass energy, as one of the main sources of renewable energy in Europe, will inevitably be one of the highest priorities. In comparison, China is clearly lacking in biomass energy development. The biomass power plants and biomass briquette fuel projects are all being developed in an unregulated and unsupported manner, with no overarching strategy or plan. Rational development of biomass resources is very important to achieving carbon neutrality in rural areas, and will facilitate agricultural modernization, increase the income of rural households, and improve the living environment in the rural areas. This energy resource definitely has great potential.

5.3.2 Method for Analyzing Spatial Distribution of Microscopic Biomass Energy Resources

1. Multispectral Signatures of Various Surface Features in Remote Sensing Image

Unlike the rural roofs identified from optical satellite images, biomass resources such as crops are hard to identify from such images. On one hand, crops and vegetation are hard to identify in optical satellite images. On the other hand, the high resolution of the images requires a huge amount of data, making it difficult to implement large-scale applications and promotions. In light of that, this publication proposes the method of using remote sensing satellite images containing multispectral information as the database for extracting biomass resources such as crops.

Remote sensing satellite image is a type of image created by remotely receiving the electromagnetic wave reflected or radiated from the surface object. The range of electromagnetic waves differs from that of conventional optical satellite images, and contains the data of various wavelengths such as visible light, near-infrared, and short-wave infrared, so such images contain more information on surface features. The commonly used remote sensing satellite images in the world are from the Landsat satellites of the United States, the SPOT satellites of the UK, and the Sentinel satellites of the European Space Agency. China has launched its domestically developed civilian high-resolution remote sensing satellites in recent years. The data collected by these satellites are used for environmental monitoring, meteorological study, resource survey, agriculture, and urban planning. In this publication, the open-source Sentinel-2 satellite images are used as the database for extracting the crops. Figure 5.15 shows the visible light-based part of the remote sensing image of various surface features in a northern region. The image contains crops, structures, other vegetation, and water bodies. With 10 m resolution, the surface features do not have clear contours, but the abundant multispectral information is enough to identify the surface features in large areas.

Through field survey and visual interpretation, a quantity of samples are collected to observe the multispectral signatures of various surface features. The study area is a double-cropping area, where wheat is sowed in winter and corn in summer, so the average spectrograms of the samples from March and August are generated, as shown in Fig. 5.16a and b. As shown in the figures, the spectral curve of the crops

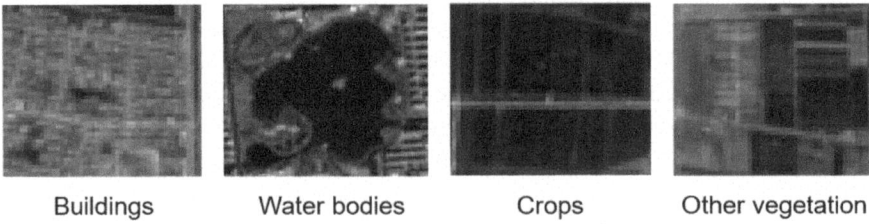

Fig. 5.15 Visible light based part of the remote sensing image of various surface features

5.3 Analysis of the Biomass Resources of the Rural Areas in China …

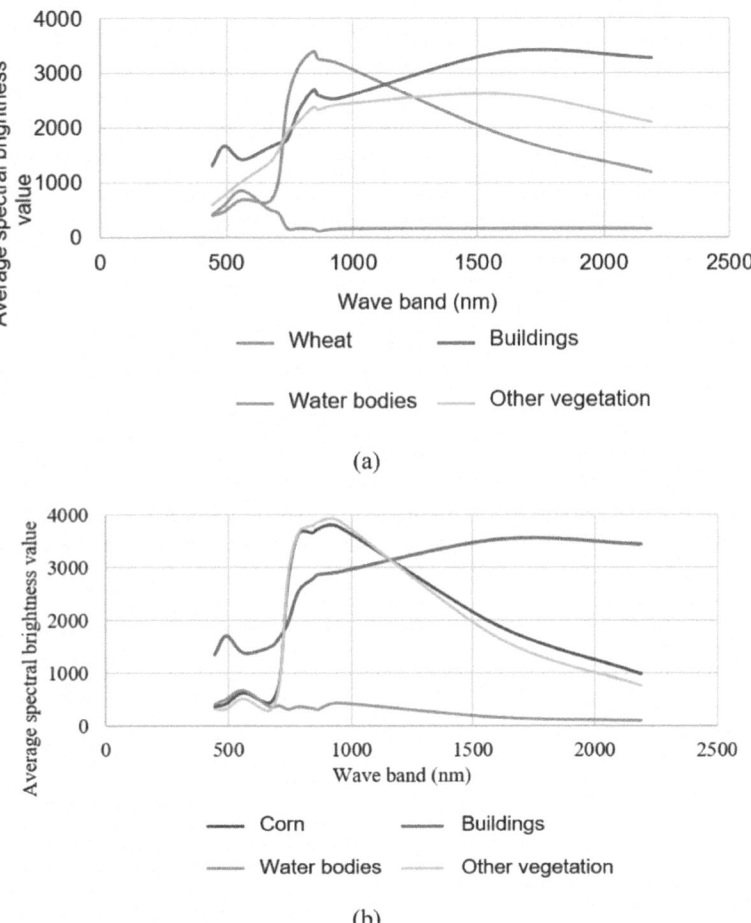

Fig. 5.16 Average spectral brightness values of various surface features sampled **a** Remote sensing image of March; **b** Remote sensing image of August

in March is highly distinguishable from the other surface features, while that of the crops in August is quite similar to the other vegetation but is highly distinguishable from the water bodies and structures.

2. Method for Extracting Crops from Remote Sensing Satellite Images of Multiple Time Phases

Remote sensing satellite images contain ample band data, but such data is not all used for identifying the crops, so the redundant information is excised by certain criteria, so as to increase calculation efficiency. The methods commonly used for vegetation identification include normalized difference vegetation index (*NDVI*), enhanced vegetation index (*EVI*), and ratio vegetation index (*RVI*). This publication uses the normalized difference vegetation index (*NDVI*). This index indicates the growth and leaf area index of crops, and is universally recognized as the index that characterizes vegetation the most effectively. *NDVI* is calculated by Eq. (5.2), where the input parameters are near-infrared band brightness value and visible red light band brightness value, and its value range is (−1,1). The closer to 1 this value is, the clearer the vegetation feature in the pixel. Figure 5.17 shows the *NDVI* value distribution diagram calculated for August in the study area. The darker the green color in the diagram, the greater the proportion of the vegetation in the pixel. The *NDVI* value of water bodies is less than 0 and is red, while the values of the structures and bare land are less than 0.5.

$$NDVI = \frac{DN_{NIR} - DN_R}{DN_{NIR} - DN_R} \qquad (5.2)$$

where DN_{NIR} is the near-infrared band brightness value;

DN_R is the visible red light band brightness value.

In the study area, two crops are planted, i.e., wheat and corn, so the crop growth pattern and *NDVI* value variation pattern are analyzed to extract the pixels that

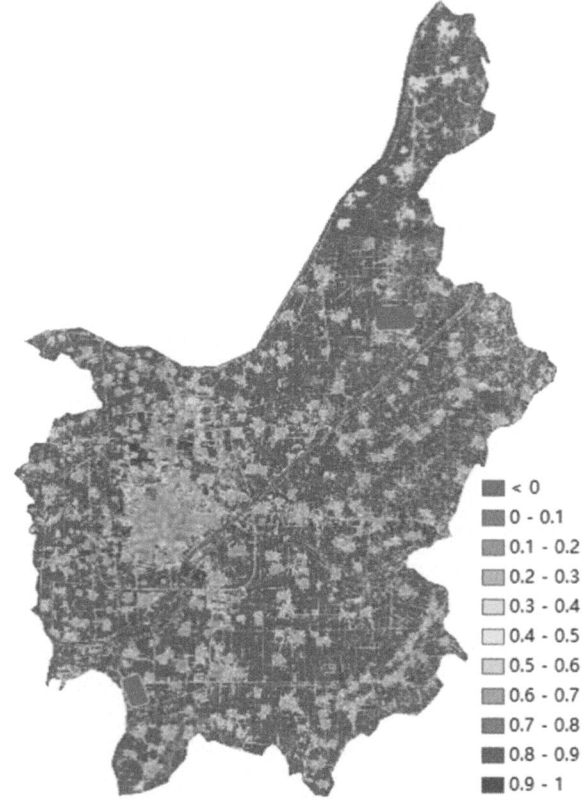

Fig. 5.17 *NDVI* value distribution diagram corresponding to the remote sensing image of August in the study area

5.3 Analysis of the Biomass Resources of the Rural Areas in China … 135

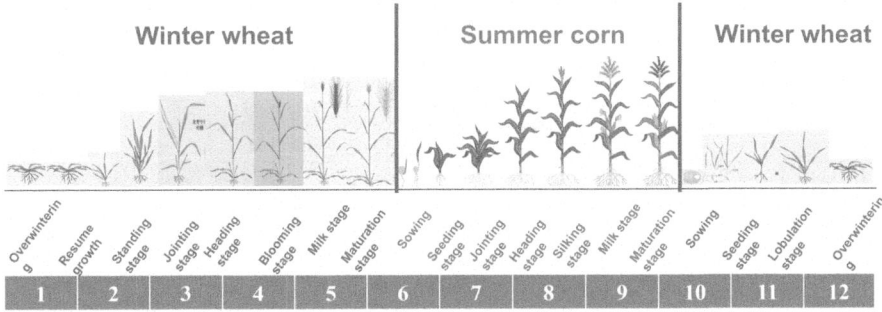

Fig. 5.18 Schematic diagram for growth cycle of summer corn and winter wheat

represent the crops. Figure 5.18 is the schematic diagram for the growth cycle of summer corn and winter wheat. It must be noted that the stages of crop growth are defined by specific criteria, but this publication does not focus on the growth pattern of the crops but instead probes into the growth conditions in certain periods so as to identify the crops, so the growth process is only briefly described.

As seen in Fig. 5.18, the sowed summer corn experiences the seeding stage and trefoil stage, where the nutrition in the seeds is depleted and seedlings grow in the soil. Then it comes to the jointing stage, where the stem grows long quickly and the leaf width increases. Then it comes to the heading stage and silking stage, where the plant grows the fastest and a large volume of biomass is accumulated. Finally, it comes to the maturation stage, where the nutrient substance of corn is quickly accumulated until it is eventually harvested. Winter wheat first experiences the germination stage and seeding stage, where the wheat seedlings grow out from the soil. Then it comes to the lobulation stage, in which the seedling age starts from the trefoil stage and ends at the elongation stage. Then it experiences the winter, when the aboveground part of the plant stops growing or grows slowly. In early spring, wheat resumes growth. Then wheat experiences the jointing stage, heading stage, and blooming stage, when wheat experiences the fastest metabolism and the biomass is accumulated the fastest. Then it comes to the maturation stage, where the grains are fully grown.

Based on the growth pattern of summer corn and winter wheat, the remote sensing images with low cloud coverage from different dates in the study area are used to calculate the *NDVI* value. For winter wheat, Fig. 5.19 shows the visible light-based remote sensing images in different growth stages and the calculated *NDVI* values. In October, which is the sowing period, the image largely shows bare land and the NDVI values are all less than 0.4. In November which corresponds to the seeding stage and lobulation stage, the land exhibits slight green, but some areas with low seedling ratio still exhibit the earthy yellow color of bare land, so a few *NDVI* values are greater than 0.5, but most of them are still less than 0.4. In January when the crops resume growth and in March when they experience the elongation stage, the crops keep growing, the green color in the image becomes darker, and the *NDVI* values of winter wheat keep increasing. The values are greater than 0.6 in the elongation stage. In the maturation stage in May, the green becomes the darkest, and most of

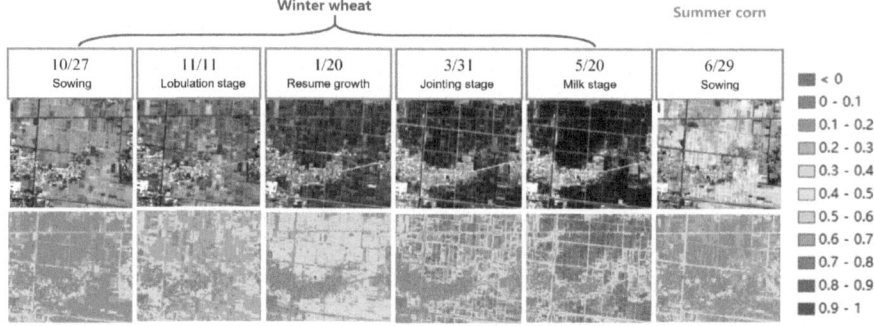

Fig. 5.19 Satellite image and NDVI image of winter wheat at main time nodes

the *NDVI* values of the wheat pixel are greater than 0.7. In June, the winter wheat has been harvested, so the pixel shows bare land again and exhibits the earthy yellow color and all the *NDVI* values are less than 0.4. It is worth noting that the other crops, the grassland and woods still exhibit vegetation features and their pixel *NDVI* values are still greater than 0.5. This exceptional scene happens because winter wheat is harvested in summer.

As seen in Fig. 5.20, the remote sensing image of summer corn in June exhibits bare land because that is the sowing stage, and the *NDVI* values of the pixels are less than 0.4. In August, which is the maturation stage, the image exhibits a dark green color, and most of the *NDVI* values are greater than 0.7, but corn cannot be visually distinguished from the other vegetation or crops. In October, corn was harvested and the image shows bare land again, and the *NDVI* values are all less than 0.4.

As described above, the growth pattern of the crops can be directly observed from the remote sensing images, and the growth pattern can be associated with the

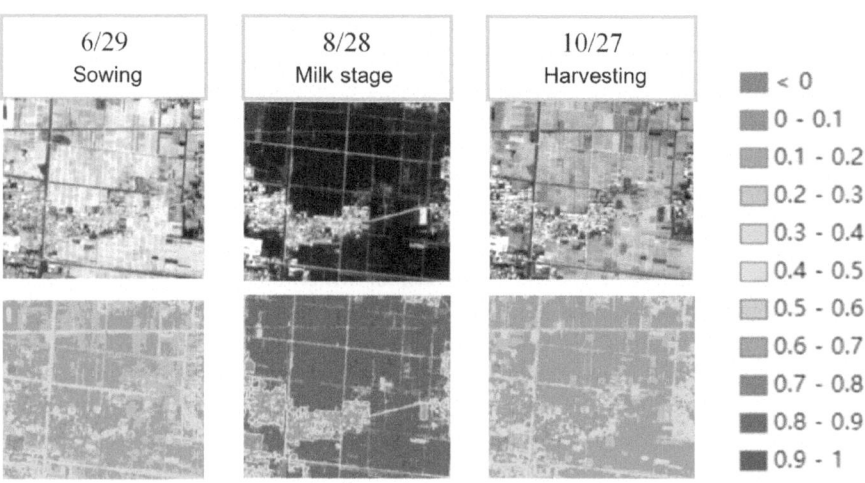

Fig. 5.20 Satellite image and NDVI image of summer corn at main time noes

5.3 Analysis of the Biomass Resources of the Rural Areas in China ...

variation of *NDVI* values, so the pixels of various crops can be extracted according to the *NDVI* values of different time phases. To extract the pixels on a large scale, the input data can be simplified, and only the *NDVI* values of the time phases that best represent the crop variation will be selected and investigated. For winter wheat, only the *NDVI* images of May and June are used for extraction. For summer corn, the *NDVI* images of June, October, and November are used for extraction, and in these months, the pixels of summer corn exhibit bare land, vegetation, and bare land. The limit of *NDVI* can be determined according to the samples, and the indexes for the aforesaid variation pattern can be established. Figure 5.21a shows the quantity of sampled *NDVI* values from three different periods of summer corn pixel. In August, as can be seen, the *NDVI* values of most pixels are greater than 0.6. In June and October, the values are all less than 0.35. Figure 5.21b shows the quantity of sampled *NDVI* values from two different periods of winter wheat. As can be seen, the *NDVI* values of most pixels are greater than 0.5 in May and they all are less than 0.35 at the end of June.

Figure 5.22 shows the quantity of the pixels that correspond to different *NDVI* values of the other vegetation, structures, and water bodies in the aforesaid 4 time periods. As can be seen, the other vegetation does not exhibit a clearly distinguishable pattern, most *NDVI* values of the structures are less than 0.5, and most *NDVI* values of the water bodies are less than 0, but these features are hard to directly distinguish according to the change of the time phases.

Based on the aforesaid variation pattern of *NDVI* values, the crop pixels can be extracted by using the decision tree method. The decision tree is one of the supervised classification methods, and it is essentially a visualization of the decision-making process. The data attributes are continuously assessed in a customized range of conditions and the branches are created. More nodes are derived from a single node to form a tree, thus representing the overall decision-making logic [24]. The variation pattern of the *NDVI* values of the aforesaid crops in different time phases is designed as the decision tree extraction rule, as indicated in Fig. 5.23. For summer corn, the *NDVI* values greater than 0.6 in August all represent vegetation, and the

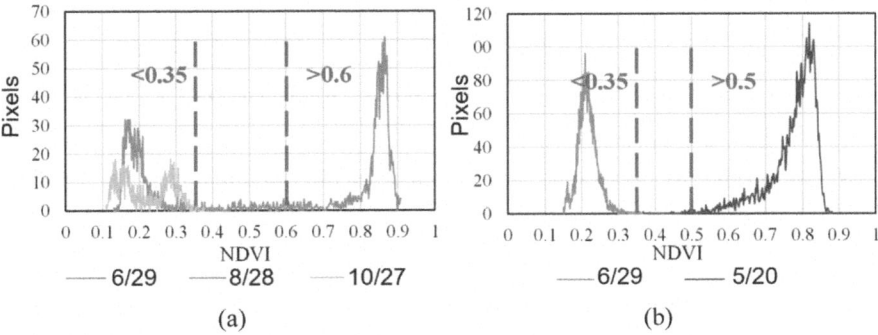

Fig. 5.21 Variation pattern of *NDVI* statistics of the crops in different time phases in the sampling areas. **a** Summer corn; **b** winter wheat

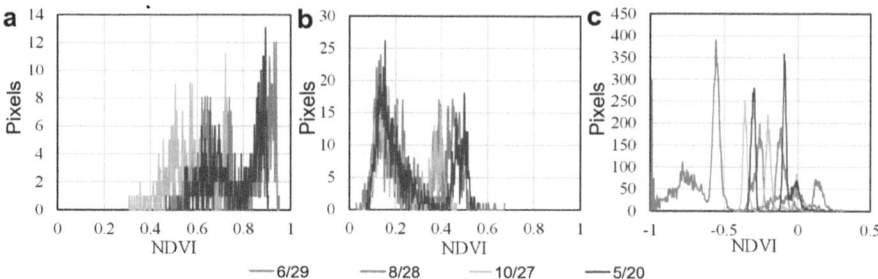

Fig. 5.22 Variation pattern of the *NDVI* statistics of the other surface features in the sampling areas. **a** Other vegetation; **b** structures; **c** water body

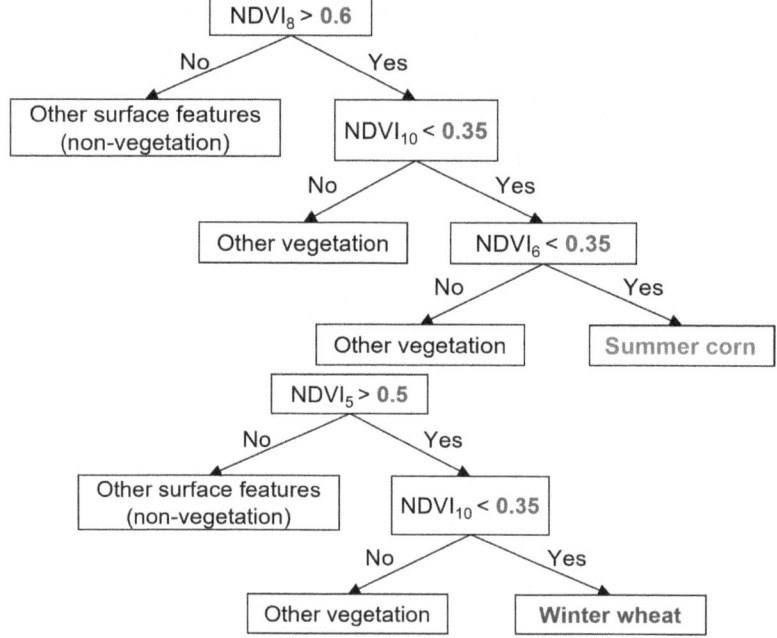

Fig. 5.23 Decision tree model used for extracting the crops in the study area

NDVI values other than that represent the other surface features. Hence, the *NDVI* values less than 0.35 in October and June all represent summer corn, and the *NDVI* values other than that all represent the other vegetation. For winter wheat, the *NDVI* values greater than 0.5 in May all represent vegetation and the values other than that represent the other surface features. Hence, the *NDVI* values less than 0.35 in June all represent winter wheat, and the values other than that all represent the other vegetation.

5.3 Analysis of the Biomass Resources of the Rural Areas in China … 139

The extraction process is designed by using the aforementioned decision tree model and is then executed, to obtain the pixels of the winter wheat and summer corn in the study area, i.e., the planting areas of these two crops, as shown in Fig. 5.24.

Figure 5.25 contains the visible light-based remote sensing satellite images of some areas in May and the winter wheat identification result. As is seen, the method proposed herein can desirably extract the planting areas of winter wheat.

The study area contains several towns. The identification results of the aforesaid model are allocated to the towns and are then compared against the planting areas determined by the governmental authorities, as shown in Table 5.4. The overall

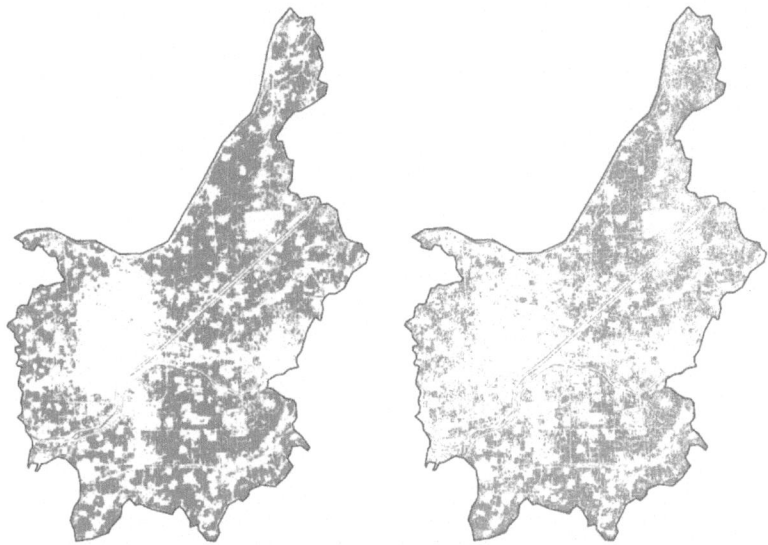

Fig. 5.24 Distribution of the crops identified by the method proposed

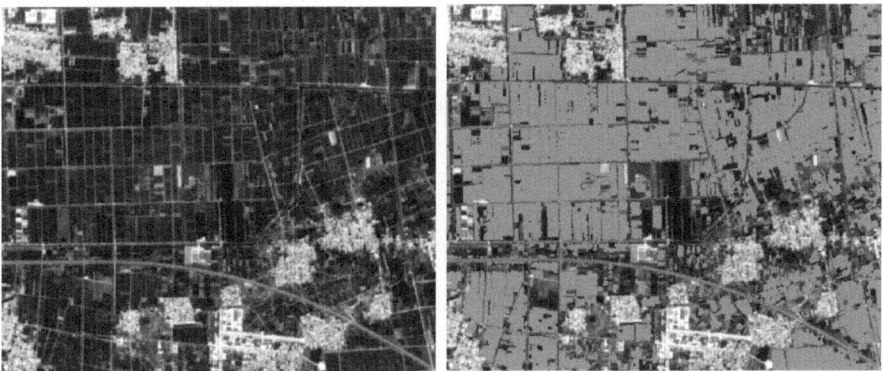

Fig. 5.25 Winter wheat extraction results of some areas

Table 5.4 Comparison of the identified area and the actual planting area of the towns in the study area

Town No	Winter wheat			Summer corn		
	Counted planting area (10,000 mu)	Identified planting area (10,000 mu)	Accuracy (%)	Counted planting area (10,000 mu)	Identified planting area (10,000 mu)	Accuracy (%)
1	2.61	2.83	91.7	1.95	2.02	96.5
2	7.27	6.78	93.2	7.15	6.35	88.7
3	5.03	4.61	91.7	4.04	3.40	84.2
4	1.86	2.17	83.6	1.83	2.51	63.1
5	3.05	2.56	83.8	2.75	2.25	81.8
6	2.21	2.06	93.1	1.68	1.60	95.2
7	4.06	4.46	90.0	2.53	3.98	42.8
8	2.98	2.77	92.9	2.85	2.75	96.6
Total	29.07	28.28	97.3	24.78	24.17	97.5

Note 1 mu666.67 m^2

identification accuracies of summer corn and winter wheat are both greater than 97%. The identification accuracies of most towns are more than 80%, but a few towns have low identification accuracies because of the problem in stitching the remote sensing images from different time phases. By now, the planting areas of various crops have been extracted from the remote sensing satellite images of multiple time phases, and the identification result contains spatial distribution information.

5.3.3 Analysis of the Potential of Macroscopic Biomass Resources

Biomass resources can be classified into crop straw, residual materials from agricultural products processing, forestry wastes, residual materials from livestock breeding, and domestic waste. To determine the total quantity of rural biomass resources in China, the total quantity and distribution of various biomass resources will be analyzed next.

1. Crop Straw

Crop straw is the most important biomass resource in China. As a traditional great agricultural country, the crop planting area of China was 2.53 billion mu (1.69×10^6 km^2) and its grain output was 683 million tons in 2021. The counted crop production output is relatively accurate, and with consideration of the crop production output of the regions published in *China Statistical Yearbook 2022* and the straw-to-grain ratio and harvestable coefficient of various crops, the crop straw resources of the provinces of China can be calculated. In 2021, the total quantity of crop straws in

5.3 Analysis of the Biomass Resources of the Rural Areas in China …

China was approximately 967 million tons, and with the utilization coefficient taken into account, this number was about 760 million tons. The resource distribution of the provinces is given in Table 5.5. The crop straws in China are mostly distributed in Northwest China, North China, and the middle and lower reaches of the Yangtze River. In terms of total quantity, Heilongjiang, Henan, and Shandong are the regions that have the highest abundance of crop straw resources, which account for 30.1% of the total quantity of China. Among all types of crops, corn, rice, and wheat are the main sources of straws, and these three crops account for 93.7% of the total straw production output.

2. Residual Materials from Agricultural Products Processing

Residual materials from agricultural products processing refer to the wastes generated when the harvested agricultural products are processed. For example, rice husk is generated when rice is processed and accounts for over 20% of the weight of rice; corncob is generated after the corn is threshed, is mainly composed of hemicellulose, cellulose, lignin and ash content, and also accounts for over 20% of the weight of corn ear; there are also the other residual materials such as peanut shell, bean dregs, and bagasse. These residual materials are mostly generated by food processing factories and sugar refineries, hence they can be easily collected and used as resources. Table 5.6 lists the proportion and production output of the residual materials generated by processing of various crops in China, and indicates that the total quantity of residual materials from agricultural products processing in China is about 160 million tons.

3. Forestry Wastes

In 2021, the forestry land area in China was about 324 million hm^2, the forest area about 220 million hm^2, the forest coverage 22.96%, and the forest stock 17.56 billion m^3. The forestry wastes are mainly generated during forest harvesting, wood processing, and branch pruning. The forestry wastes are mainly found in the Northeast and Southwest of China, and the annual production output is about 240 million tons.

4. Residual Materials from Livestock Breeding

Residual materials from livestock breeding refer to the excrement of the livestock and the washing water. These biomass resources, if not properly disposed of, will cause environmental pollution, and the bacteria and residual drugs will impact the health of the residents. After the treatment processes such as anaerobic fermentation, these materials can be used as biomass energy resources. As livestock breeding becomes more and more centralized, the excrement of the livestock experiences innocent treatment processes such as anaerobic fermentation at the livestock farms or energy companies and thus is converted into usable biomass energy. According to the second nationwide pollutants survey and the results of the related studies, the total quantity of livestock excrement generated in China every year is about 3.05 billion tons (wet weight), which is equivalent to about 110 million tons of standard coal. The top 3 provinces with the largest production are Sichuan, Henan, and Yunnan.

Table 5.5 Harvestable quantity of different types of crop straws in some of the provinces (autonomous regions or municipalities) (in 10,000 t)

Province (autonomous region and municipality directly under the Central Government)	Rice	Corn	Wheat	Beans	Potatoes	Other cereals	Total
Beijing municipality	0.1	44.1	5.7	0.4	0.7	0.3	51.3
Tianjin	38.7	180.1	61.0	0.5	0.7	1.9	282.8
Hebei province	35.1	3146.5	1240.2	19.9	100.6	47.1	4589.4
Shanxi province	1.3	1488.3	205.5	19.1	47.7	66.2	1828.1
Inner Mongolia autonomous region	81.5	4558.4	132.7	164.2	90.8	156.1	5183.7
Liaoning province	313.0	3287.3	0.6	25.6	13.7	34.8	3675.1
Jilin province	504.8	5235.1	0.9	59.6	13.0	45.4	5858.8
Heilongjiang province	2148.0	6791.4	15.4	696.6	16.4	10.3	9678.1
Shanghai municipality	82.8	1.3	6.4	0.1	0.1	0.4	91.1
Jiangsu province	1930.6	541.4	1166.9	66.0	22.4	16.8	3744.1
Zhejiang province	456.3	46.2	42.0	28.2	36.8	3.3	612.8
Anhui province	1547.1	1222.2	1477.7	90.2	15.3	4.5	4357.1
Fujian province	316.8	17.8	0.0	7.4	86.6	1.4	429.9
Jiangxi province	2017.5	39.3	2.8	31.1	50.0	0.9	2141.7
Shandong province	68.9	3942.3	2225.9	48.3	79.6	7.7	6372.7
Henan province	336.5	3123.5	3210.3	69.2	87.7	8.8	6836.0
Hubei province	1832.4	583.6	347.2	39.4	93.7	4.0	2900.2
Hunan province	2610.1	422.3	6.8	38.8	85.6	5.3	3168.9
Guangdong province	889.7	70.6	0.1	6.9	105.7	0.4	1073.5

(continued)

5.3 Analysis of the Biomass Resources of the Rural Areas in China …

Table 5.5 (continued)

Province (autonomous region and municipality directly under the Central Government)	Rice	Corn	Wheat	Beans	Potatoes	Other cereals	Total
Guangxi Zhuang autonomous region	820.0	331.3	0.5	16.2	54.7	2.6	1225.3
Hainan province	102.4	0.0	0.0	0.8	18.1	0.0	121.3
Chongqing municipality	374.7	289.0	5.0	24.8	221.7	5.2	920.5
Sichuan province	1135.0	1231.4	202.4	84.4	428.6	38.0	3119.8
Guizhou province	317.2	291.3	27.5	20.6	232.4	33.5	922.5
Yunnan province	373.8	1126.8	51.1	64.2	165.3	40.2	1821.3
Xizang autonomous region	0.3	3.4	16.6	1.1	0.3	54.6	76.3
Shaanxi province	0.0	804.8	329.0	19.2	82.7	28.7	1264.4
Gansu province	0.0	860.1	216.7	21.5	175.4	40.1	1313.8
Qinghai province	0.0	20.5	30.0	2.1	23.6	18.4	94.6
Ningxia Hui Autonomous region	0.0	352.3	14.7	0.6	28.3	6.7	402.6
Xinjiang Uygur Autonomous region	0.0	1354.6	495.8	6.5	16.3	8.5	1881.7
Nationwide	18,334.6	41,407.2	11,537.6	1673.2	2394.6	692.2	76,039.4

5. Domestic Waste in Rural Areas

Domestic waste is disposed of in three ways, i.e., landfilling, incineration, and composting. At present, domestic waste in China is mostly disposed of through sanitary landfill, which will occupy large areas of land and will easily cause secondary pollution. As the waste incineration industry has progressed, the proportion of waste incineration in harmless disposal has increased gradually, and there is great potential

Table 5.6 Proportion and production output of residual materials generated by processing of various crops in China

Types of crops	Residual materials	Proportion of residual materials in weight (%)	Production output of crop (10,000 t)	Residual material production output (10,000 t)
Corn	Corncob	>20	27,255	5451
Rice	Rice husk	>20	21,284	4257
Beans	Bean dregs	15–20	1966	393
Sugarcane	Bagasse	50	10,666	5333
Peanut	Peanut shell	35	1831	641
Total	–		63,002	16,075

for using domestic waste as an energy resource. The per-capita organic domestic waste production in rural areas is 0.6 kg/(person·day), so the annual total production is about 109 million tons.

The aforesaid biomass resources are summarized in Table 5.7. If the calorific value of various biomass resources is converted into standard coal, the annual potential total quantity of rural biomass energy is about 736 million tce. The crop straws take up the highest proportion, which is about 50.6%. The other biomass resources, ranked in terms of proportion, are forestry wastes, residual materials from livestock breeding, residual materials from agricultural product processing, and rural domestic waste.

Table 5.7 Summary of rural biomass resources in China

Types of biomass resources	Annual production (100 million t)	Equivalent standard coal (100 million t)	Proportion (%)
Crop straw	7.60	3.72	50.6
Residual materials from agricultural product processing	1.61	0.80	10.9
Forestry wastes	2.40	1.44	19.6
Residual materials from livestock breeding	30.50	1.10	14.9
Domestic waste in rural areas	1.09	0.30	4.0
Total	–	7.36	100

5.4 Total Quantity and Development Potential of Rural Hydropower Resources in China

5.4.1 Overview of Rural Hydropower

A rural hydropower station in China is defined as a small hydropower station with $\leq 50{,}000$ kW installed capacity. In terms of their social attributes, these facilities refer to the rural infrastructure and public facilities that are built and managed by the local governments, that serve the agriculture, rural areas, and farmers, and that provide electrical power to support the socioeconomic sustainable development of the rural areas and counties. Such power stations include on-grid small hydropower stations, off-grid small hydropower stations, and miniature hydropower stations. The rural hydropower stations are characterized by moderate scale, less investment, short construction duration, and fast generated profits, so they can easily motivate many parties to participate, and they can be developed by the country, local authorities, groups, enterprises, and even individuals.

Rural hydropower resources exist in many locations, cover large areas, have massive total quantities, account for 39% of the total exploitable hydropower resources, and play an important part in electric power structure adjustment. However, the rural hydropower resources are distributed unevenly, because they are mostly found in the upper reaches of the Yangtze River, the upper reaches of the Pearl River, and the upper and middle reaches of the Yellow River. These are the areas in West China inhabited by ethnic minorities and the border areas. These areas are sparsely populated and have scattered loads, so they are hard to be covered by large power grids and are unsuitable for long-distance power transmission and supply. These areas also contain many nature reserves and are the key areas subject to water and soil loss control. Rural hydropower projects are characterized by scattered development, local grid connection, nearby power supply, and low power generation costs. The development of rural hydropower in these areas and the implementation of small hydropower fuel replacement projects are beneficial supplements to the large power grid, and are also an important way to protect and improve the ecological environment.

1. Installed Capacity of Rural Hydropower Stations

By the end of 2020, China had 43,957 rural hydropower stations, and their total installed capacity was 81,338,000 kW, which accounted for 22.0% of the total installed capacity of the hydropower system of China and 3.7% of the total installed capacity of the electrical power system of China [25]. Figure 5.26 shows the changes in the number and total installed capacity of rural hydropower stations in China in 2011–2020.

In recent years, the priority in rural hydropower projects has shifted from construction to management, where efforts are made to remedy the shortcomings of the hydropower projects and enhance industrial supervision. Green modification and development of hydropower projects are the inevitable paths to meet the requirements of the era and to support the sustainable development of the hydropower stations.

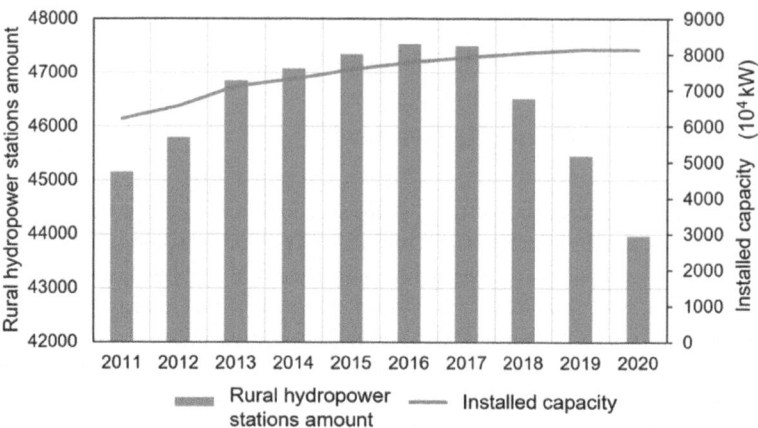

Fig. 5.26 Number and total installed capacity of the rural hydropower stations in China in 2011–2020

In 2020, the number of rural hydropower stations increased by 70 and the additional installed capacity of operational power generation equipment was 807,000 kW, including a newly added installed capacity of 576,000 kW and an added installed capacity of 231,000 kW achieved by the modified power generation equipment. The newly added installed capacity was mainly concentrated in Southwest China. In 2020, the added installed capacity of rural hydropower projects with 10,000 (incl.)–50,000 kW (incl.) scale accounted for 66.5% of the added total installed capacity, that of the projects with 1,000–10,000 kW scale accounted for 29.1%, and that of the projects with below 1,000 kW scale accounted for 4.4%.

By the end of 2020, the number of counties in China that had rural hydropower stations was 1,529, which was 13 less than in 2019, and these counties were mostly in Southwest China, Central China, and South China. The first three provinces ranked in this regard were Sichuan, Yunnan, and Guangxi. The installed capacity and annual energy production of the rural hydropower projects in the regions are listed in Table 5.8.

2. Energy Production of Rural Hydropower Projects and Local Power Consumption

In 2020, the energy production of the rural hydropower projects in the country was 242.37 billion kWh in total, which was 10.95 billion kWh less than in 2019, with a 4.3% decrease year-on-year (YoY). The annual average utilization hours of rural hydropower were 2,980 h, which was 130 h less than in 2019, with a 4.2% decrease YoY. Based on the coal consumption in the power generation industry, the energy production of the rural hydropower projects in 2020 was equivalent to reduced consumption of 74 million tce, reduced CO_2 emission of 185 million tons, and reduced SO_2 emission of 973,000 tons. In 2020, the energy production of the rural hydropower projects in the country accounted for 17.9% of the energy production of all hydropower projects and 3.2% of the total energy production of the country,

5.4 Total Quantity and Development Potential of Rural Hydropower ...

Table 5.8 Installed capacity and annual energy production of the hydropower projects in the regions [24]

Region	Installed capacity by the end of the year (10,000 kW)	Annual energy production (100 million kWh)	Region	Installed capacity by the end of the year (10,000 kW)	Annual energy production (100 million kWh)
Beijing municipality	2.2	0.1	Hunan province	631.5	194.8
Tianjin	0.6	0.1	Guangdong province	773.3	157.7
Hebei province	40.5	5.7	Guangxi Zhuang autonomous region	464.7	144.0
Shanxi province	20.9	3.1	Hainan province	49.9	11.1
Inner mongolia autonomous region	9.5	2.1	Chongqing municipality	302.0	87.1
Liaoning province	47.1	9.7	Sichuan province	1147.5	440.5
Jilin province	62.0	19.1	Guizhou province	369.5	132.6
Heilongjiang province	40.3	14.0	Yunnan province	1267.3	384.1
Jiangsu province	3.8	0.5	Xizang autonomous region	42.2	10.7
Zhejiang province	411.5	82.0	Shaanxi province	172.9	52.9
Anhui province	114.9	28.8	Gansu province	287.9	120.3
Fujian province	729.9	164.2	Qinghai province	108.4	56.4
Jiangxi province	351.0	89.1	Ningxia Hui autonomous region	0.6	0.1
Shandong province	8.6	0.7	Xinjiang Uygur autonomous region	174.8	65.8

(continued)

Table 5.8 (continued)

Region	Installed capacity by the end of the year (10,000 kW)	Annual energy production (100 million kWh)	Region	Installed capacity by the end of the year (10,000 kW)	Annual energy production (100 million kWh)
Henan province	51.2	8.5	Xinjiang production and construction corps	43.1	14.6
Hubei province	391.9	119.9	Under the direct leadership of ministries	12.2	3.7
Total				8133.8	2423.7

which was 0.3 percentage points lower than the previous year. In 2020, 232.15 billion kWh of electricity from rural hydropower projects was diverted to the power grid and 4.44 billion kWh was supplied to local users, which accounted for about 1.9% of the energy production.

5.4.2 Green Transformation of Rural Hydropower

China ranks first place in the world in terms of installed capacity and energy production of small rural hydropower projects. However, some of the small hydropower projects in some regions are unregulated, developed excessively, and devoid of effective supervision, which has led to ecological environmental problems such as water flow reduction or disappearance in some river reaches and ecological degradation. In 2017, the Ministry of Water Resources started the endeavor of developing green small hydropower stations, and guided the small hydropower industry to reform its development model and transition to a sustainable development model that is friendly to the ecological environment. In December 2018, the 4 authorities including the Ministry of Water Resources, the National Development and Reform Commission, the Ministry of Ecology and Environment, and the National Energy Administration jointly issued the *Opinions on Clearing and Rectification of the Small Hydropower Projects in the Yangtze River Economic Belt*, which initiated the efforts to clear and rectify the small hydropower projects in the Yangtze River Basin. By the end of 2021, over 3,500 power stations in the Yangtze River Basin had been demolished, and 25,000 had been rectified. This facilitated the green transformation of the small hydropower projects in the Yangtze River Basin. Regarding the Yellow River Basin, the 7 authorities including the Ministry of Water Resources jointly issued the *Notice* on *Clearing and Rectification of the Yellow River Basin* in 2021. This notice stipulated that the small hydropower projects in the county-level administrative regions along the trunk stream and tributaries of the Yellow River in the 8 provinces including Qinghai, Gansu, Ningxia, Inner Mongolia, Shaanxi, Shanxi, Henan, and Shandong must be cleared and rectified by the end of 2024. In December 2021, the 7 authorities including the Ministry of Water Resources jointly issued the *Opinions on Further Rectification of Small Hydropower Projects*, which explicitly stated that the endeavor must follow the principles of "compliance with laws and regulations, being practical and realistic, making policies suitable for a specific project, and ensuring safety", and the prominent conflicts and interests in demolition and rectification of the small hydropower projects must be addressed properly, so as to promote the transformation, upgrading and green development of the small hydropower projects. This document clarified the requirements for aggressive and prudent clearing and rectification of the small hydropower projects in a period henceforward [26].

In recent years, the regions responded to the national policy of developing pilot green small hydropower stations by vigorously building pilot green small hydropower stations. By the end of 2021, 870 pilot green small hydropower stations had been built in China. These projects came as great examples and substantially contributed to the green transformation of the small hydropower projects in the country.

5.4.3 Development Potential of Rural Hydropower in China

China has numerous rivers with abundant resources for small hydropower projects. According to the *National Rural Hydraulic Energy Survey and Assessment Report* (2008), the Chinese mainland has 128 million kW of technically exploitable resources for small hydropower projects with an individual installed capacity of $\leq 50,000$ kW and the energy production is 535 billion kWh, which makes China rank the first place in the world in this regard. These resources are found in 1,715 mountainous counties in 30 provinces (regions or municipalities). The technically exploitable quantity of East China is 22.84 million kW, which is 18% of the national total, that of Central China is 25.67 million kW, which is 20% of the national total, and that of West China is 79.53 million kW, which is 62% of the national total. The 6 provinces (autonomous regions or municipalities) in Southwest China including Sichuan, Guizhou, Yunnan, Xizang, Guangxi, and Chongqing possess the most abundant resources for small hydropower projects, which is 61.93 million kW, accounting for 48.4% of the national total. The 6 provinces (autonomous regions or municipalities) in Northwest China including Inner Mongolia, Shaanxi, Gansu, Ningxia, Qinghai, and Xinjiang have relatively concentrated resources for small hydropower projects, which is 17.6 million kW, accounting for 13.7% of the national total. The resources for small hydropower projects in Northeast China are mostly found in Jilin and Heilongjiang and the quantity is 5.5 million kW, accounting for 4.3% of the national total. The resources for small hydropower projects in Central China are mostly found in Hunan, Hubei, and Jiangxi and the quantity is 20.78 million kW, accounting for 16.3% of the national total. The resources for small hydropower projects in East China are mostly found in Zhejiang, Fujian, and Guangdong and the quantity is 22.17 million kW, accounting for 17.3% of the national total.

The total energy production potential of the rural hydropower projects in China is 535 billion kWh, which is equivalent to about 161 million tce, considering the current coal consumption index in the electrical power industry is 301.5 gce/kWh.

5.5 Summary

The rural areas in China possess abundant distributed zero-carbon energy in the form of rooftop PV, biomass, and rural hydropower, but such resources have not been adequately exploited. Under the carbon neutrality goal, the rural areas should make greater efforts to develop the aforesaid resources, so as to change the rural areas from consumers of fossil energy to producers of zero-carbon energy, to contribute to the carbon neutrality endeavor of China, improve the living quality and economic income of the rural dwellers and to improve the environmental quality. This chapter proposes the research methods based on the total quantity and spatial distribution of rooftop PV, biomass energy, and rural hydropower resources, and presents the calculation results. The following conclusions are obtained:

(1) For the rural rooftop PV resources, at the microscopic level, satellite images are used as basic data, and the corrected U-net neural network model has been created and applied. With consideration of PV panel installation methods, the potential total quantity of household-level and village-level rooftop PV resources is calculated and the spatial distribution of the resources is determined. At the macroscopic level, the contours of the villages across the country have been identified by using low-resolution images. Then typical regions are sampled from the provinces, and the rural roofs in these regions are identified by using high-resolution satellite images, so as to calculate the proportion coefficient of the rural roofs and the rural roof area in the provinces. Then the potential total installed capacity of rural rooftop PV systems in China is calculated as 2 billion kW and the annual power generation potential is 2.9 PWh.

(2) For the biomass resources represented by crop straws, the quantity of different types of biomass energy have been identified by using remote sensing satellite images of multiple time phases and the statistics of the relevant departments. As calculated, the annual total quantity of rural biomass energy in China is about equivalent to 736 million tce. The crop straws account for about 50.6%. The other resources, ranked in descending order, are forestry wastes, residual materials from livestock breeding, residual materials from agricultural product processing, and rural domestic wastes.

(3) The current installed capacity of the rural hydropower projects in China is about 81,338,000 kW and the annual energy production is 242.37 billion kWh. In terms of resource potential, the potential installed capacity of the rural hydropower

projects in China is about 128 million kW and the annual energy production potential is about 535 billion kWh.

Open Access This chapter is licensed under the terms of the Creative Commons Attribution-NonCommercial-NoDerivatives 4.0 International License (http://creativecommons.org/licenses/by-nc-nd/4.0/), which permits any noncommercial use, sharing, distribution and reproduction in any medium or format, as long as you give appropriate credit to the original author(s) and the source, provide a link to the Creative Commons license and indicate if you modified the licensed material. You do not have permission under this license to share adapted material derived from this chapter or parts of it.

The images or other third party material in this chapter are included in the chapter's Creative Commons license, unless indicated otherwise in a credit line to the material. If material is not included in the chapter's Creative Commons license and your intended use is not permitted by statutory regulation or exceeds the permitted use, you will need to obtain permission directly from the copyright holder.

Chapter 6
How Rural Areas Can Take the Lead in Transforming from Energy Consumers into Energy Prosumers

In the context of dual carbon goals, the role of rural areas in the energy system of China will change significantly, and one of its key characteristics is that rural areas will transform from traditional energy consumers into new energy prosumers in the future. This hinges on the resource endowment and load characteristics of rural areas. Rural areas are rich in renewable resources and do not have a high demand for their own energy, so they have the natural conditions to realize the abovementioned transformation. However, changing the role of rural areas in the energy system necessitates a reasonable top-level design and technical route planning. First, the ways of energy use in rural areas are complex and diverse. Compared with cities, rural areas now have a low electrification rate and rely heavily on fossil energy for heating or cooking. Besides, agricultural machinery and vehicles in rural areas also rely heavily on gasoline, diesel, etc. To radically change the current situation of energy use in rural areas, taking active measures to guide the comprehensive electrification of energy use in rural areas is the core. Second, although there is a large amount of photovoltaic (PV) installed capacity in rural areas, PV power has strong volatility and the power demand in rural areas is stochastic, resulting in a mismatch between PV power supply and electrical load. Moreover, transformers in rural areas have limited capacity, making it difficult for them to carry huge PV potential, and the unordered connection of a large amount of photovoltaics to a power grid will bring challenges to the dispatching of the power grid and even threaten the safe operation of the power grid.

Starting from the construction of a new distributed power system of "photovoltaic, energy storage, direct current and flexibility (PEDF)" in rural areas, the full electrification of rural areas, and the commercialization of rural biomass fuels, this chapter explores the technical routes of transforming rural areas from energy consumers into energy prosumers and further discusses and analyzes relevant issues.

6.1 New Rural PEDF Power System

Although rural areas have abundant rooftop PV resources, the traditional development mode of PV in rural areas is far from satisfactory. The operating mode of the traditional PV system in rural areas is as follows: A developer rents the rooftops of peasant households to install PV panels, PV power is directly inverted and fed to a power grid, and peasant households still need to purchase power from the power grid to meet their own power demand. This mode has three main problems: ① Unfairness. Because the transformer capacity is limited, only a few "swift-footed" peasant households in each village can enjoy the on-grid capacity, and it does not benefit all the villagers. ② Inadequacy. As power generation and power utilization are two systems, the developer obtains profits from power sales, but farmers do not truly benefit. ③ Imbalance. PV power is stochastic and volatile, and the direct connection of it to a power grid is likely to generate garbage electricity, making it impossible to utilize the energy storage resources in rural areas and leading to a waste of a large amount of peak shaving power resources.

In recent years, the installed capacity of residential distributed generation photovoltaic (DG PV) in China has grown rapidly. In 2022, the installed on-grid capacity of residential DG PV in China increased by 16.9% compared to the previous year. The installed on-grid capacity of residential DG PV in some provinces of China as of the first half of 2023 is shown in Fig. 6.1 [27]. As can be seen from the figure, currently the total installed capacity of residential PV in China reaches 95.02 GW, accounting for about 4.8% of the installed potential of rooftop PV in rural areas of China. The residential installed capacity in Shandong is the highest, about 23.22 GW, and only accounts for 7.3% of the installed potential of rooftop PV in rural areas of the province. The vast majority of rural rooftops are still undeveloped. However, DG PV has affected the dispatching of the local power grid and even led to a "negative feed-in tariff". This is because PV power is strongly volatile and stochastic, which not only makes it difficult for the dispatching center to determine the power generation characteristics of DG PV but also causes the grid power flows to change. Most of all, transformers in rural areas have limited capacity, making it difficult for them to accommodate more PV power to be fed into a power grid. Take a village with 100 households in Shandong as an example. The total capacity of transformers in the village is about 500 kVA, and the installed PV capacity for each RRB is about 20 kW. Due to the safety limit on the transformer capacity, the installed PV capacity in the village can be 400 kVA (80% of transformer capacity) according to the current rooftop PV operating mode in rural areas. This can only meet the rooftop PV capacity for 20 RRBs and only accounts for 20% of the total PV capacity of the village, while most of the rooftop PV capacity is wasted. Therefore, the transformation of the energy role of rural areas requires reasonable top-level design and technical scheme; besides, with the rapid extension of residential DG PV, rural areas are in urgent need of a new type of PV power system to guide their energy transition.

As a result, the traditional PV system cannot guide the decarbonization of rural energy use, and rural areas need a new type of power system that allows farmers

6.1 New Rural PEDF Power System

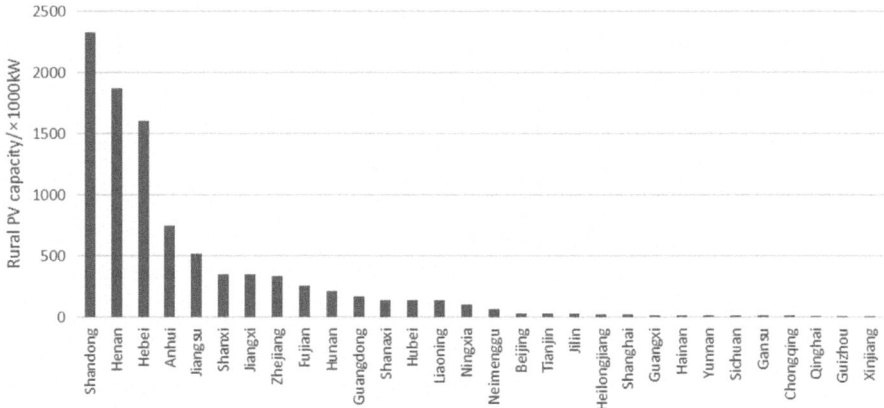

Fig. 6.1 Installed capacity of residential DG PV in some provinces (autonomous regions and municipalities directly under the Central Government) (as of the first half of 2023). *Data Source* Official website of the National Energy Administration of China

to enjoy PV power and brings real benefits to them. In this context, the new rural PEDF power system emerges [28]. For the so-called PEDF power system, "P" mainly refers to the power supply by renewable energy dominated by RRB rooftop PV; "E" refers to multiple energy storage systems, including heat storage equipment, electric vehicles, electric agricultural machinery and other electrochemical energy storage equipment; "D" refers to the use of a DC system, which, for one thing, is to reduce losses in power transmission and conversion, and for another, is to facilitate the adoption of low-cost and controller-free control based on busbar voltage signals; "F" refers to the flexible operation of the power system, which is the ultimate purpose. The power system may change its operating state according to user requirements or grid dispatching signals to adjust its power generation and energy consumption curves. To overcome the problems of unfairness, inadequacy, and imbalance brought by the traditional rural PV system, the basic principle of the new PEDF power system is to prioritize self-generated power for self-use and orderly feed surplus power to a grid, as shown in Fig. 6.2. The operating mode of this system is as follows: The PV power is used preferentially to meet the power demand of peasant households and the charging demand of power storage equipment, and the surplus power is orderly fed into a grid as dispatched. Compared with a traditional rural PV system, such a system has 3 advantages: ① The transformer capacity expansion is less stressful. Because a considerable proportion of PV power is consumed locally, the peak power of on-grid electricity is much less than that of electricity directly inverted and fed to a power grid, and more peasant households are allowed to join this system; ② The energy storage advantages of rural areas are fully utilized to make PV power no longer garbage electricity and add no dispatching pressure on the power grid; ③ Peasant households use PV power directly, which not only allows farmers to receive economic benefits but also realizes the zero-carbon energy use in rural areas and drives the full electrification of rural areas.

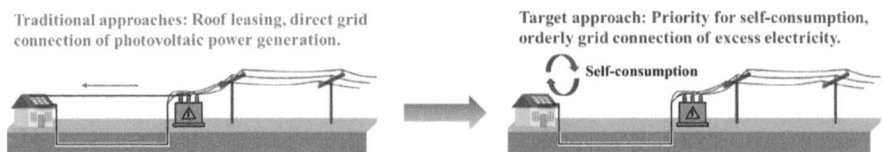

Fig. 6.2 Comparison of two types of PV power systems

6.1.1 Technical Characteristics of New Rural PEDF Power System

The topological graph of the new rural PEDF power system is illustrated in Fig. 6.3, and the system consists of a household unit and a public unit. The household unit includes 10–20 kW or above PV installed on the rooftop, an indoor DC distribution system, and low-voltage DC busbars connecting various electrical appliances, batteries, household DC charging points, etc. Household appliances mainly include lighting equipment, air conditioners, TVs, computers, electric cooking appliances, electric water heaters, electric bicycles, small electric agricultural machinery, and electric heating installations. Currently, most of the DC appliances listed above are available on the market, and the DC conversion of household appliances or tools can be achieved through simple modifications to existing AC equipment. Relying on 10–20 kW PV panels, the annual energy production of each household can reach 10,000–20,000 kWh, which can meet all domestic electricity demand of a rural household, including winter heating for the main room of 20–60 m^2 in northern China, heating and air conditioning in southern China, and charging of various vehicles and production tools. The surplus power in spring, summer, and autumn can be used for agricultural production and the production and processing of agricultural by-products in the village. Moreover, it can respond to the dispatch of the power grid through busbar voltage control and be fed to the power grid at the appropriate time. Indoor batteries can meet the power consumption demand of night lighting and household appliances. If it is overcast for several days, the power demand of basic lighting and household appliances can be further met by relying on the power storage in batteries and electric equipment (e.g., electric vehicles, electric agricultural machinery, etc.) batteries. After the rooftop PV power of peasant households meets the indoor and charging equipment demands, the surplus power is orderly fed into the village-level busbar.

About 25 RRBs with a rooftop PV system can form a transformer coverage area, and 4 such areas can form a new rural power system for a village with about 100 households. The household DC microgrids are connected to the DC busbar of the village via the DC/DC (DC to DC) converters between households, and this DC busbar connects to the DC microgrids of all households and public units. A public unit mainly includes PV installed in public idle space, PV systems of public buildings in the village (such as the Villagers' Committee, the Party-Masses Activity Center, and the collective factory building of the village), public DC charging points,

6.1 New Rural PEDF Power System

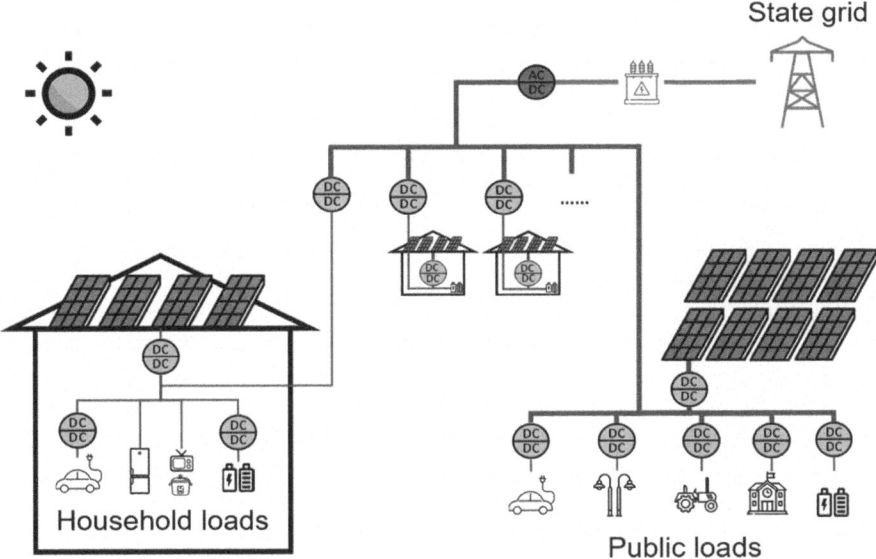

Fig. 6.3 Topological graph of the new rural PEDF power system

village-level centralized power storage devices (e.g. storage modules and large agricultural machinery batteries), and various electrical facilities for production (pumps, agricultural by-product processing equipment, etc.). The village-level DC microgrid connects to the microgrids of all households. It also connects to public power generation and storage units to supply power for night lighting and various public facilities in public areas. It is converted into AC power through inverters, stepped up by transformers, and then connected to an external power grid. The village-level DC microgrid can enable the mutual complementation of electricity between households and provide energy for public and production facilities. All households can sell and feed their surplus power to a power grid within the period required by the power grid after it is allocated uniformly through the village-level DC microgrid and the public batteries, thus obtaining higher economic income. The power storage capacity of household and public batteries and the batteries of electric vehicles and electric agricultural machinery may approach the total amount of electricity generated in a day by the PV cell connected to the village-level DC microgrid. Through such integration, the PV power is converted from an unadjustable and uncontrollable fluctuating power supply into an adjustable and controllable high-quality power supply, which can meet the power demand in a sunless period and also can be fed to a power grid within the specified period according to the requirements of the power grid. This new rural power system has been demonstrated at the village level and will be promoted and implemented in different regions.

To treat the development of rooftop PV as a foundation of the new energy revolution aiming for full electrification in rural areas, priority needs to be given to addressing the energy demands of living, production, and transportation in rural areas and replacing the current coal, fuel, gas, and direct straw burning in rural areas. Peasant households have different mechanized and transportation equipment due to the different agricultural production they are engaged in. With the full electrification of these agricultural machineries and equipment, the battery resources owned by peasant households will be more than those of urban residents. Adopting the battery swap method with a standard modular battery, each household can own 60 kWh of energy storage resources. As the utilization rate of most agricultural machinery and equipment is not high, these batteries usually can be placed in the battery cabinet of each household as distributed energy storage resources. By relying on rooftop PV and 60 kWh batteries, various domestic energy demands, such as cooking, heating, domestic hot water, lighting, and household appliances, can be well met. Even if it is overcast for 3 days, various power demands can be guaranteed provided that power storage is pre-planned. One-fourth of the 20 kW PV energy production can meet the domestic power demand of a peasant household, and the remaining three-fourths can be used for partial production and on-grid electricity sales. The 60 kWh power storage capacity can fully meet the demand of night lighting, life, entertainment, etc., for power supply. As to the power for production use (such as irrigation and processing of agricultural by-products), the demand response method can be implemented, and production can be organized according to the power supply condition. In this way, finally, the remaining 1/2–2/3 of the power (average 12,000–16,000 kWh/year per household) can be fed and transmitted to a power grid by using the existing rural power grid system, thus becoming a new source of economic income. If the average distribution capacity per household is 5 kW, the annual on-grid power transmission time will be 2,500–3,000 h, which is twice the time the power generated directly by PV is connected to a power grid. When the 60 kWh battery resources are used, power can be transmitted during the peak electrical load period instead of the PV power generation period. This allows rural PV to take part in the regulation of the large power grid, thus making PV turn from "garbage electricity" to regulated power. In this case, the rural rooftop PV system in the mode of "integration of production, consumption, regulation and storage" shall be developed, and the renovation of power systems and installation of rooftop PV systems should be carried out at the village level using the transformer station as a unit. For each transformer station of a DC microgrid, the "output without input" of electricity shall be realized; when multiple low-voltage transformer stations under a high-voltage transformer station have realized DC transformation, the "output without input" of electricity can be realized at the upper-level high-voltage transformer station; however, for a village-level power grid without DC transformation, the original mode, i.e., the "input without output" of electricity, shall be still implemented, with the schematic diagram shown in Fig. 6.4. In this way, the unidirectional flow of current at each transformer station can be realized, thus avoiding the change in power flows and the impact on the power grid as a result of unordered connection of electricity to and disconnection from the power grid.

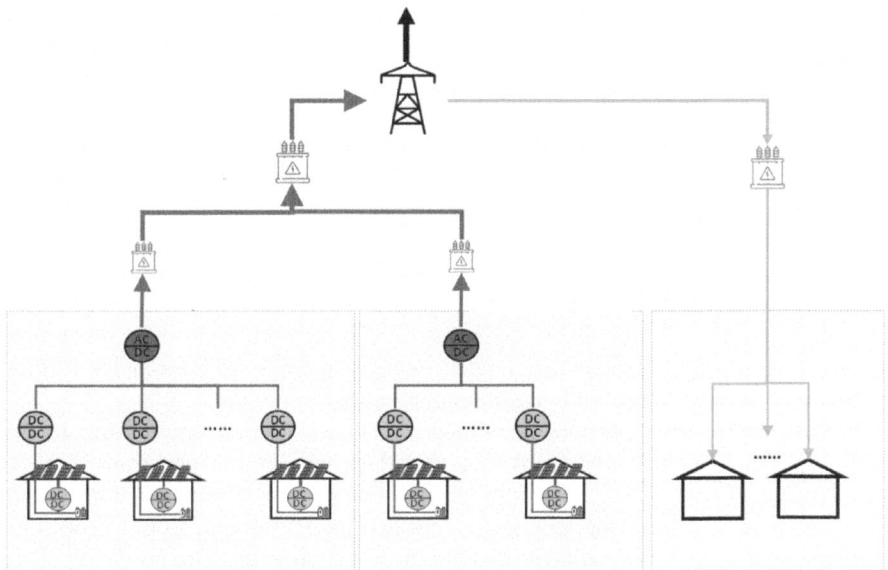

Fig. 6.4 Schematic diagram of the new rural PEDF power system

6.1.2 Economic Analysis of New PEDF Power System

The cost and benefit of a typical rooftop PV system are analyzed by taking the construction of a system with an installed PV capacity of 20 kW by a single household as an example, and the initial investment in equipment is approximately RMB 100,000. In the cost structure, PV panels and supports account for about half of the system investment, and DC/DC conversion components make up 34% of the system investment; in addition, there is also investment in engineering materials, regulation, metering, etc. No initial investment subsidy is considered for the indoor investment portion. The peasant household independently obtains a low-interest loan through the green finance or rural revitalization project and owns the property rights. In this analysis, a 2% PV low-interest loan is considered, and the equal principal repayment and interest payment method is adopted for its repayment. If the Villagers' Committee or the peasant household invests in the system, the peasant household can utilize electric power free of charge; it can also transmit the surplus power to a power grid at the right time and use the on-grid electricity sales revenue to repay the principal and interest of the investment. Suppose the annual average energy production time is 1,200 h, then the total annual energy production of rooftop PV of a single peasant household can reach 24,000 kWh. Considering that the total power consumption of production, living, and transportation is 10,000 kWh, there are still 14,000 kWh that can be transmitted to the power grid. There is no need to consider solar curtailment because the on-grid electricity of the rooftop PV-micro-grid has regulation and storage properties. By feeding the stored electricity to the power grid in the set period, the on-grid electricity selling price can be increased to RMB 0.55/ kWh, and the annual electricity sales revenue can reach RMB 7,700. According to this mode, if the peasant household only bears the investment in the indoor power system and the renovation cost of the transformer coverage area system is borne by the power grid company or the Villagers' Committee, the investment in the indoor portion can be returned in about 15 years. In recent years, the costs of PV modules and converters are still falling, and both state and local subsidies are provided for rural PV, energy storage, rural grid renovation, clean heating, and other projects. If all peasant households can enjoy these policy subsidies, each household can reduce investment by about RMB 20,000 on average and the payback period can be further shortened. In fact, with the technological advancement and large-scale production and application of PV panels and related devices, there is also room for further reduction of average household investment amount in the future.

Apart from the indoor portion, the village-level public portion requires investment in the capacity expansion and renovation of village-level power grid transformers, the village-level centralized energy storage equipment, the power connection system (PCS) modules in transformer coverage areas and the PV mounted on the rooftops of public buildings in the village, and the average investment required is RMB 10 million/natural village, covering 300 households/village. The investment mainly includes PV panels, transformers, DC microgrids, and batteries. It is suggested that this portion of investment should be made by the relevant departments of the state after collecting existing subsidies related to rural energy and made as energy infrastructure projects (such as special rural infrastructure upgrading and renovation, rural revitalization fund, and rural power grid upgrading and renovation projects). With the constant evolution of electrified agricultural machinery, the manufacturing cost of the energy storage equipment in the public part of the village-level PEDF new power system will be substantially reduced, further reducing the investment cost of the system. Each village-level power grid can be operated and maintained by 1–2 professional technicians.

6.1.3 Analysis of Popularization Mode of PEDF New Power System

Analysis of the roles of the participants in the PEDF new power system is shown in Fig. 6.5. There are four participants, i.e., government, grid company, investor, and rural household. These roles can be overlapped, because one party can play multiple roles. The investor is the leading party in the DG PV sector. An investment entity can be an individual or an enterprise. The model in which an individual is the investment entity is quite simple, for the project has low complexity, and fewer interested parties are involved. In the model in which an enterprise is an investor, in comparison, the project is usually complex, a large amount of investment is required, and a large number of governmental departments and financial institutions have to be

6.1 New Rural PEDF Power System

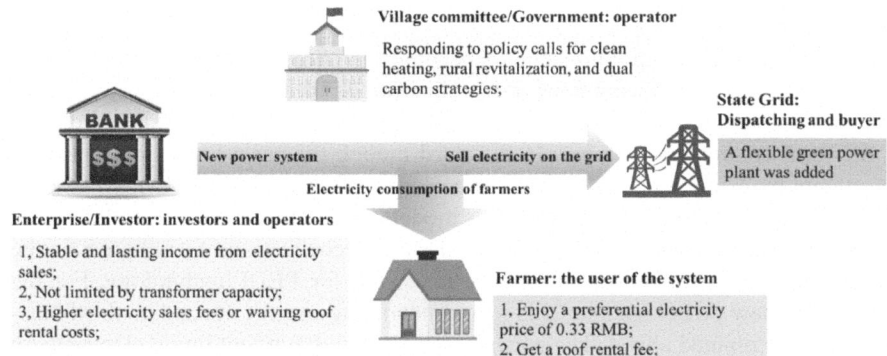

Fig. 6.5 Analysis of the roles of the participants in rural PEDF new power system

coordinated in the operation process. The rural household is usually the consumer, one of the beneficiaries of the distributed energy, the end user of the electric energy, and also the main entity creating the profits in this business model. The financial institution mainly provides or raises the project capital, and it is the source of funds for the DG PV project; it is also the decisive factor in the effectiveness of funds turnover and normal operation of the project in this business model. The government enacts the relevant policies and industrial standards, monitors the market, approves and grants various subsidies, enacts the regulations to regulate the behaviors of the market players, resolves the disputes and conflicts in the market, and creates a benign market environment for DG PV. The grid company power diverts the DG PV electricity to the power grid and organizes electric power transmission. Currently, the grid company is also responsible for the measurement and settlement of electricity quantity, electricity fees, and subsidies, and plays an important part in facilitating the development of DG PV.

The Ministry of Agriculture and Rural Affairs and National Administration for Rural Revitalization promulgated the *Notice on Issuing the Guidelines for Social Capital Investment in Agriculture and Rural Areas (2021)* [29], which stipulates that innovative investment and financing models will be applied according to the reality of the agriculture and rural areas in various regions, the approaches such as sole proprietorship, joint venture, joint operation and leasing, and the business models such as franchised project, government-invested and individual-run project and individual-invested and government-assisted project will be explored to improve the incentives for agricultural promotion and to facilitate rural revitalization steadily and prudently. According to the national policy, three business models are being used in the rural PV market. In the first model, a large energy company provides investment and rents the roofs of rural households, and the contractual relation and settlement procedure are quite simple. In this model, however, the rural households only participate to a small extent and gain a low proportion of benefit, the enterprise bears high operation and maintenance costs in the later stage, and the investor is usually a large state-owned enterprise. This model has the longest payback period among all models. In the second model, a collectively owned enterprise provides the investment and the rural households become shareholders by providing funds or mortgaging the land management right. This model, to a certain extent, is dependent on the managerial competency of the village cadres and the basic conditions of the rural collective economy. In the third model, the rural households provide investment by using loans, use the electric energy generated by them, and divert the surplus electricity to the power grid. This model has the least overall social cost and shortest payback period, best aligns with the full electrification of rural areas in the future, and is an application scenario better suited for PEDF technology. For the rural PEDF new power system, it is recommended to use the second or third business model explained above. The general idea is "output without input, electricity in and by rural households". "Output without input" means all the electricity needed by the villagers is generated by the village's PV power generation system, and the surplus electricity can be transmitted out. "Electricity in and by rural households" means the villagers will be the owners of the energy infrastructure in the village. For the equipment in the rural households, the farmers will provide an investment of their own volition. The funds can be sourced from green finance loans with low or zero interest. For the public equipment in the village, the collectively owned enterprise will provide investment, become the integrated operator of clean energy, invest in and operate the local bulk coal management projects, conduct the maintenance and services related to distributed renewable energy, produce biomass briquette fuel, manage rural energy stations and undertake other clean energy development efforts. The government will purchase the services and the users pay the fees, to ensure the projects generate enough investment income. The government will assist the projects to obtain policy-supported green loans. To sum up, the reasonable investment method should involve the government, enterprises, and rural households, with the government playing the leader role (assisting with loan applications or participating in partial investment), enterprises playing the investor role (investing a large proportion), and rural households playing the participant role (to improve their economic benefits and enthusiasm).

6.1.4 Impact of New Rural PEDF Power System on Energy System Structure

With their PV potential greater than the power demand, rural areas will transform from energy consumers to energy prosumers. The new rural PEDF power system can not only solve the energy demand of rural areas but also transmit excess zero-carbon power to surrounding cities to meet urban residents' production and domestic electricity demand. Therefore, the new rural PEDF power system will have an important impact on the national energy system. The energy production (which can be divided into renewable energy production and thermal energy production), power consumption, and power trading situations in the regions of China in 2022 are shown in

6.1 New Rural PEDF Power System

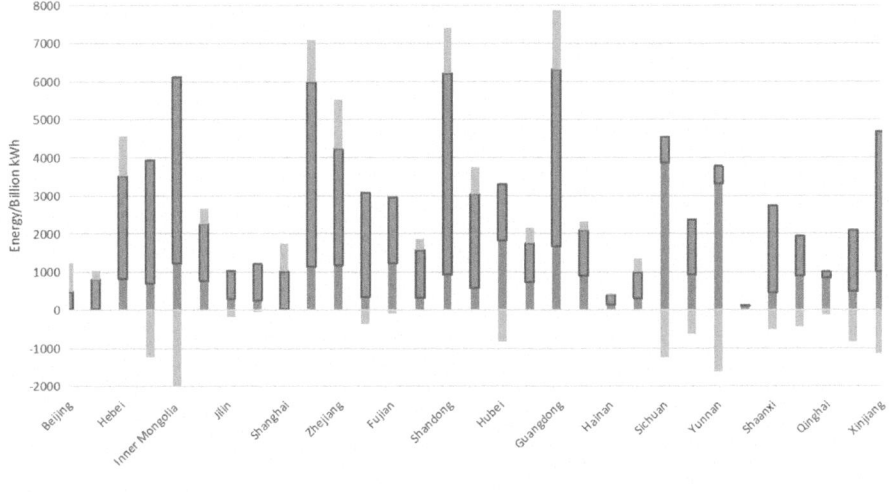

Fig. 6.6 Energy structures in some regions in China in 2022*. *Note* The data is from the official website of the National Energy Administration

Fig. 6.6. Among them, a positive power trading volume with other provinces indicates purchasing power from other provinces while a negative one indicates selling power to other provinces. It can be seen from the figure that the energy production in Shanxi, Inner Mongolia, Xinjiang, Sichuan, Yunnan, and Hubei was greater than their own power demand. The difference was that hydropower accounted for a large proportion of the energy production in Sichuan and Yunnan while thermal power accounted for a large proportion in Shanxi, Inner Mongolia, and Xinjiang. Beijing, Jiangsu, Zhejiang, Guangdong, Shandong, etc., needed to purchase power from other provinces due to large power demand. Except for Sichuan, Yunnan, Qinghai, and Hubei, energy demand in other provinces was mainly met with thermal power. Such an energy structure is to the disadvantage of achieving the carbon neutrality goal.

The impact of the new rural PEDF power system on energy structures in various provinces is shown in Fig. 6.7. The rural PV power can potentially replace about 2.9 PWh of thermal power in the country. In terms of the total quantity, the said potential is equivalent to 49% of the current thermal power generation in the country, and such potential is greater than thermal energy production in Hebei, Henan, Sichuan, Yunnan, Xizang, and Qinghai. To achieve the carbon neutrality goal in China's power system in the future, a PV penetration rate of 30% is needed. According to the analysis in Chap. 4, with a vast roof area, rooftop PV power generation is entirely possible to become the main method of PV power generation in rural areas in the future. Therefore, the new rural PEDF power system will play a vital role in China's achievement of carbon neutrality.

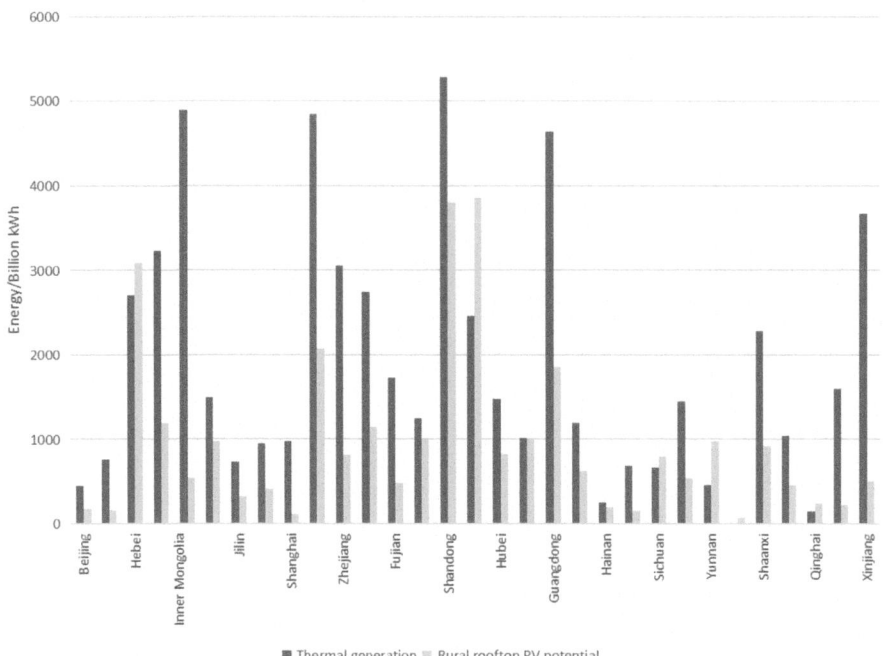

Fig. 6.7 Comparison of rural rooftop PV power generation potential and thermal energy production in various regions

6.1.5 Policy Suggestions for the New PEDF Power System

Existing policies provide support for improving the economy of the new rural PEDF power system, however, their guidance and motivation for the construction of the new power system are insufficient. More targeted policies should be introduced to help promote the new power system.

1. Tariff

Tariff is currently the main factor restricting the new rural PEDF power system. With the income of the power system from power sales, increasing the tariff will significantly improve the economy of the power system. However, the tariff is directly related to people's livelihood, and simply increasing the tariff will affect rural households' quality of life. Therefore, the dynamic time-of-use tariff can be considered to increase the feed-in tariff at night, which will help improve the economy of energy storage and constitute a positive feedback reward for systems that can flexibly respond to grid dispatching. In the meantime, rural households can be guided to increase their power consumption during peak periods to help promote local consumption of PV power. Furthermore, the share of carbon trading can also be increased in power trading, to increase the economy of renewable electricity.

6.1 New Rural PEDF Power System

2. Trading Platform

The construction of distributed generation trading platforms can be accelerated so that rural households that invest in installing PV systems can sell power to nearby households with consumption loads. Unlike the diverse building types and complex metering and charging of power consumption in urban areas, each rural household has the natural production-marketing integration attribute. Therefore, allowing rural households to participate in system maintenance in real-time and monitor the sales of power to nearby households can completely solve the predicament that "rural households look on and do nothing while enterprises can't afford operation and maintenance" in the current household PV system in rural areas.

3. Subsidy

The economic comparison shows that the higher converter cost of the new power system is the biggest contributor to the higher investment cost of the current new power system compared to the direct inverter system. Technically, the difficulty of a DC/DC converter is lower than that of an inverter because a DC/DC converter does require consideration of the frequency, phase, etc., of the AC system. However, the price of DC/DC converters is currently more than twice that of inverters due to the immaturity of the market, which requires the introduction of related policies to guide the market. Therefore, the introduction of related policies is suggested, to subsidize DC converters. For the same reason, the market price of DC electric apparatus is much higher than that of AC electric apparatus, therefore, policy subsidy support is also needed for DC electric apparatus. Subsidy support for such DC devices can increase market demand, thereby promoting the industrial upgrading and development of DC devices. Similarly, subsidies for potential energy storage equipment in rural areas, such as electric vehicles and electric agricultural machinery, will also contribute positively to improving the economy of the new rural PEDF power system. It is important to accelerate the electrification of various electric agricultural and forestry machinery and implements and vehicles, promote the R&D and promotion of related new products, adjust subsidy policies for agricultural and forestry equipment, and transform the current support policies for agricultural and forestry production mechanization into support for the fuel-to-electricity projects. The task of electrification of agricultural and forestry equipment can be included in China's agricultural and forestry equipment development plan. The adoption of standard modular batteries and the use of agricultural machinery and vehicles with swappable batteries can bring into full play the role of batteries. With the households' standard module batteries becoming the key to the system, it is suggested that the state provide special financial subsidies for 50% of electric vehicle purchase costs, to comprehensively promote the construction of the rural new energy system and the electrification of agricultural machinery.

4. Guidance

Based on the village as a unit, all the resources with conditions for paving PV panels, such as rural households' rooftops, courtyards, agricultural facilities, forest land, and wasteland, can be uniformly planned, registration and "one village, one policy" can be launched by reference to the working method in the fight against poverty, demands for electric agricultural machinery and implements, electric vehicles, rural household heating and household appliance upgrading can be comprehensively considered, the PEDF method can be adopted, and construction can be conducted uniformly, in phases and batches. Developing in coordination with the PV system, the PEDF system will become an important driver of the promotion of electric vehicles and electric agricultural machinery in rural areas. It is suggested that villagers' committees cooperate by providing relevant materials for building piles, to ensure the construction of slow charging piles; public fast charging piles included in the scope of rural infrastructure and constructed with funds directly appropriated by the central or local government; and the review of DG PV systems strengthened, PV grid-connected specifications formulated, such as strictly stipulating parameters including the energy storage ratio, system flexibility, response speed and self-consumption rate, and the quality of DG PV systems improved through mandatory standards. Furthermore, in some scenarios, such as RRBs, industrial plants, and schools with a sufficient rooftop area, the scheme of "prioritizing self-generation and self-use and connecting excess power to the grid" can be promoted.

Current projects and policies related to rural energy can be integrated and financial resources, policies, and mechanisms can be centralized to fully support the establishment of the rural new energy system, including ① clean heating projects which should be PV-based electrification solutions; ② rural grid capacity expansion projects which should target at effectively collecting surplus power from the rooftop PV power in rural areas; ③ home appliances going to the countryside projects which should adjust the project content based on the characteristic of the comprehensive electrification in rural areas; and ④ agricultural mechanization projects which should advance the comprehensive electrification of agricultural machinery to make electric agricultural machinery also an important energy storage facility in the rural power system.

6.2 Electrification Solutions for Rural Energy Use in the Future

An essential condition for rural areas to transform from energy consumers to energy prosumers is the comprehensive electrification of the energy consumption side in rural areas, in which case PV power generated from rural rooftops can directly transform the energy use structure in rural areas and realize zero-carbon energy use there. Furthermore, an important goal of the new rural PEDF power system is to promote the priority consumption of PV power. Therefore, the comprehensive electrification of rural energy use is a future development trend and a requirement for constructing the new rural PEDF power system.

6.2.1 Current Situation of Electrification of Energy Use in Rural Areas

Based on data from the *China Rural Statistical Yearbook 2020* by the National Bureau of Statistics on end-use energy consumption in agriculture and rural residents' lives, the electrification rates of agriculture and rural residents' lives were calculated to be 35.2% in 2020.

The electrification rates of agriculture and rural residents' lives had significant differences among provinces. Figure 6.8 shows the electrification rates of agriculture and rural residents' lives in some regions in 2020, with those with higher rates mainly in eastern and southern China. The electrification rates of agriculture and rural residents' life in Fujian, Guangxi, Jiangsu, Hainan, Anhui, Ningxia, Beijing, Shandong, and Guangdong exceeded 40%, while the figures in northeastern and northwestern China were significantly lower than those in eastern and southern China.

The level of electrification in agriculture and rural residents' lives has improved rapidly in recent years, showing a clear growth trend, and the accelerated improvement in rural electricity access has promoted the progress of electrification in rural areas. It can be seen from Fig. 6.9 that the electrification rates of agriculture and rural residents' lives cumulatively increased by 9.2% from 2016 to 2020, showing rapid growth. The electrification rates of agriculture and rural residents' lives in four regions, i.e., eastern, central, northwestern, and northeastern China, from 2018 to 2020 are shown in Fig. 6.10. There are significant differences in levels of electrification in agriculture and rural residents' lives in different regions of China. The level of

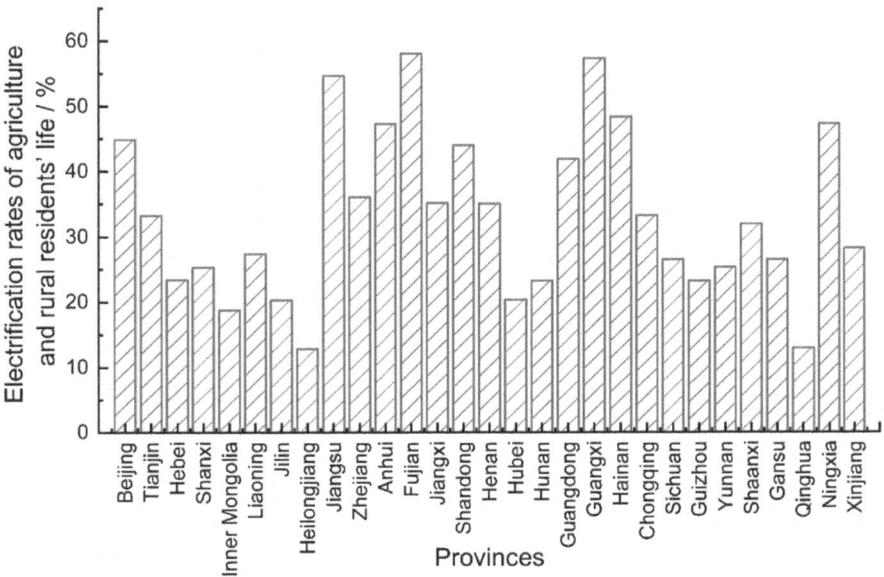

Fig. 6.8 Electrification rates of agriculture and rural residents' life in different regions of China in 2020. *Note* Relevant statistical data for Xizang, Hong Kong, Macao, and Taiwan are missing

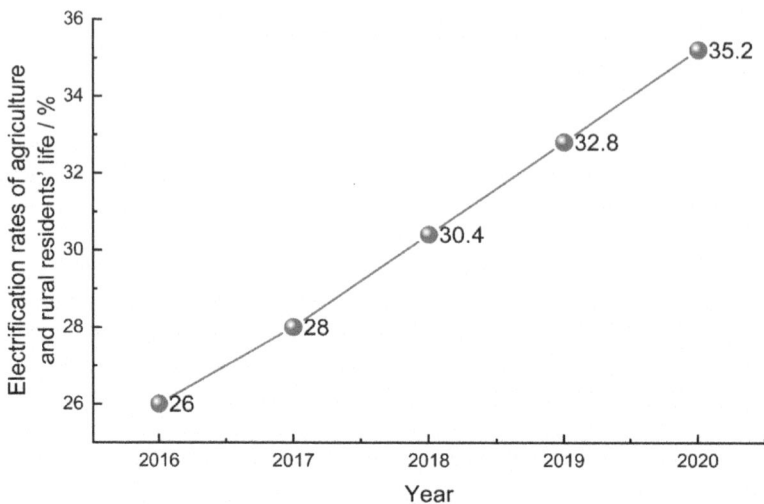

Fig. 6.9 Changes in the electrification rates of agriculture and rural residents' life in China from 2016 to 2020

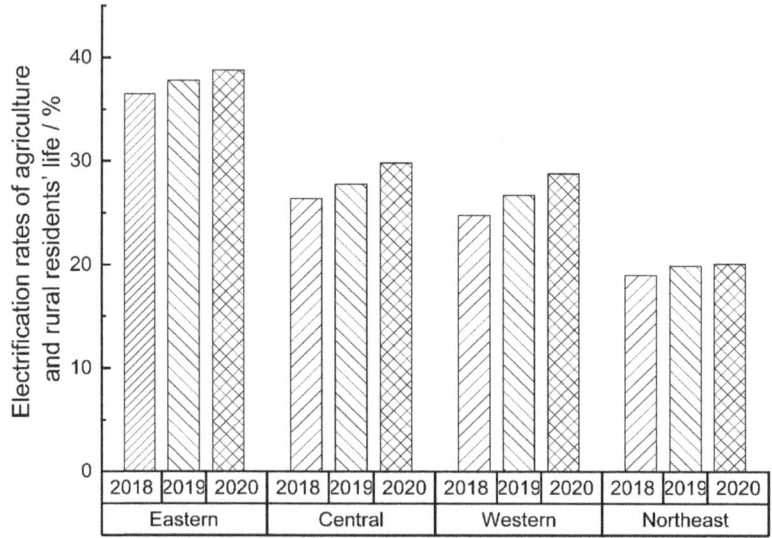

Fig. 6.10 Electrification rates of agriculture and rural residents' life in the four major regions of China from 2018 to 2020

electrification in agriculture and rural residents' lives in eastern China is significantly higher than the national average level, while the electrification level in both central and western China is being improved rapidly. According to the data for 2020, the electrification rate for agriculture and rural residents in eastern China reached 38.8%, and that in central, western, and northeast China were 29.8%, 28.8%, and 20.1%, respectively. Compared with those in 2019, the rural electrification rates in eastern, central, western, and northeastern China increased by 1.0%, 2.0%, 2.1%, and 0.2%, respectively. The continuous improvement of infrastructure in eastern China and the rapid development of new energy projects, such as DG PV power generation projects, have greatly promoted the electrification level of agriculture and rural areas. The rural electrification level in central and western China is gradually approaching that in eastern China. The energy sources the agriculture and rural residents in northeast China rely on mainly include gasoline, diesel, and raw coal, resulting in a significant gap in the rural electrification level between northeast China and other regions. From 2018 to 2020, the electrification rates of agriculture and rural residents' lives in eastern, central, western, and northeastern China cumulatively increased by 2.3%, 3.4%, 4.0%, and 1.1%, respectively. With the large-scale development of renewable energy generation projects and the continuous improvement of power infrastructure and that of rural residents' power consumption levels, agricultural power consumption and rural residents' power consumption have achieved rapid growth in central and western China, with the growth rate in these two regions significantly higher than that in other regions.

Household appliances of rural residents are increasingly diversified (Fig. 6.11). The quantity of refrigerators, air conditioners, and washing machines owned by rural residents has been increasing yearly from 2000 to 2020, and in recent years, the quantity of air conditioners and refrigerators owned by rural residents has shown a rapid growth trend. In 2020, the quantity of refrigerators, air conditioners, and washing machines owned by every 100 rural households was 103.5 sets, 89.0 sets, and 96.1 sets, respectively. This is mainly due to the development of the rural economy and the improvement of rural residents' income levels, which enables rural residents to purchase more home appliances to improve their quality of life. In recent years, some household appliances such as TVs and computers, which are used both for entertainment and life, have gradually shown a trend of stabilization or even decline. The reason is that with the popularity of the Internet and mobile phones, there have been new entertainment devices as substitutes. Overall, although there is still a significant gap in the number of home appliances owned by rural households compared with urban households, with the continuous increase in both income and consumption of rural residents, the rural household appliance market will continue to expand, and there will still be a considerable room for growth in the number of home appliances owned by rural residents. Therefore, there is still a great potential for improvement in the electrification level in rural areas.

Although there are differences in levels of electrification in agriculture and rural residents' lives among different regions, overall, the electrification levels have maintained a relatively strong growth momentum. With the deepening of the rural revitalization strategy and the continuous improvement of rural residents' income, the popularity and richness of household appliances in rural areas will be increased continuously. However, the development of electrification in rural areas is still relatively weak in terms of heating, cooking, and agricultural production. To further improve

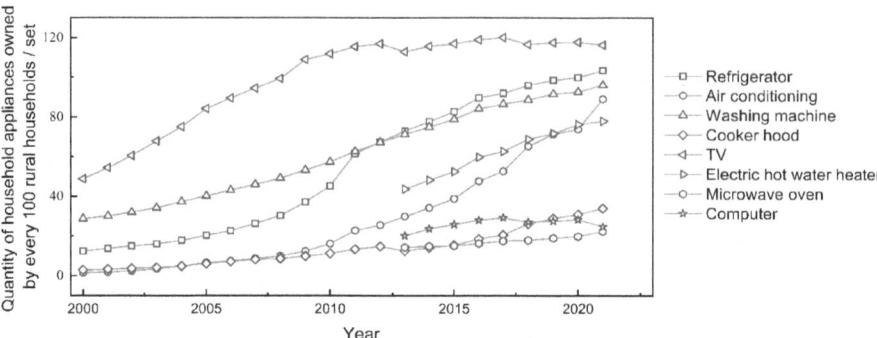

Fig. 6.11 Quantity of household appliances owned by every 100 rural households

the electrification level in rural areas, it is necessary to accelerate the development of electrification in the fields of heating, cooking, agricultural production, etc.

6.2.2 Electrification in Heating

The traditional heating method of burning bulk coals in rural areas of northern China has caused emissions of a large number of pollutants and CO_2, seriously threatening the health of residents. The electrification in heating in rural areas can significantly reduce the emissions of various pollutants and become an effective way for use of photovoltaic power, which is conducive to promoting the "self-production and self-consumption" of rooftop PV and reducing the capacity expansion pressure of transformers in rural areas.

From the perspective of terminal electricity demands, the mainstream electric heating equipment in the current market is mainly divided into two categories: direct electric heating equipment and heat pump products. Direct electric heating equipment, as the name suggests, refers to devices that generate heat directly from electricity, such as electric heaters and graphene sheets. Their advantages include simple equipment, small size, and low price, while their disadvantage is low energy efficiency. Electric storage heaters are also applied in some rural areas, because they can use thermal-storage materials such as refractory bricks to store and transfer the heat, enabling the equipment to store the heat during the valley period of power consumption, and use the stored heat to maintain the temperature inside RRBs and meet heating needs during the peak period of power consumption. This operating mode can reduce the operating costs of the equipment. Common heat pump products include air-to-air heat pumps and air-source heat pump water heaters. The advantage of such products is high energy efficiency, and the disadvantage is that the price is relatively expensive. With the implementation of China's "coal-to-electricity" policy, air-to-air heat pumps have become more popular in rural areas. Due to their advantages of instant start and stop and fast heating speed, they are suitable for the heating

6.2 Electrification Solutions for Rural Energy Use in the Future

needs of "partial-time, partial-space" in rural areas, and have achieved good results in use.

From the perspective of power sources, current electric heating can be divided into the municipal power supply and PV power supply. For municipal power supply, it is necessary to solve the problem of improving the energy efficiency of electrical equipment, i.e., using less electricity to provide more heat, which can reduce the energy consumption for building heating, decrease heating-related carbon emissions, and save heating costs. For PV power supply equipment, it is necessary to solve the problem of mismatch between the fluctuation of PV power supply and the heating demand. The commonly used methods include the combination of energy storage equipment or the use of control technology to improve the stability of PV heating. However, the current challenge faced by the equipment is how to reduce the solar curtailment rate of PV and achieve economic energy storage while improving the self-satisfaction rate of PV. It is worth mentioning that if municipal power supply is continued, the rural energy system will still assume the role of a consumer. Rural areas will become energy prosumers only when rooftop PV heating is realized in rural areas. Therefore, the use of PV power supply equipment will be the future development trend to achieve electrification in heating in rural areas. In this chapter of this book, we will introduce the technical solutions for achieving PV heating.

6.2.3 Evaluation on Electrification in Energy Consumption for Living in Rural Areas—Taking Cooking Energy as an Example

According to more than 400 questionnaires collected by a research team from Tsinghua University in provinces such as Liaoning, Shanxi, Gansu, Heilongjiang, Jiangsu, and Hainan from 2022 to 2023, the types of energy used for cooking in rural areas of China are diverse, and the energy sources mainly involve electricity, coal, firewood, liquefied gas, and natural gas. With the non-use of bulk coals in rural areas and the continuous improvement of rural power grids in China, as well as the improvement of safety and convenience of electric cooking equipment, electricity has gradually become the main choice for energy consumption for rural cooking. The increasing use of electricity reflects the improvement of living standards and the transformation of energy in rural areas. Besides, due to their convenience in use, liquefied gas tanks and natural gas are still widely used in rural cooking. Furthermore, coal and firewood are widely used in many regions due to their low cost, easy availability, and requirements for certain traditional cooking methods.

There is a situation of multiple types of energy use for cooking in rural areas of China. Some rural kitchens often "adopt several cooking methods simultaneously" in energy consumption, which may be due to ① diversity of energy supply: The energy supply in rural areas is relatively limited, and thus residents may choose multiple energy sources to meet their cooking needs based on the availability and cost of energy. ② Differences in functional requirements: Different cooking requirements may require different energy sources. For example, coal and firewood are more commonly used in traditional stoves, while electricity and liquefied gas are more suitable for modern induction cookers and gas stoves. ③ Living habits and traditional factors: There are some traditional cooking methods in rural areas, and thus residents may continue to use traditional energy sources to meet their habitual needs, such as coal and firewood. In summary, in terms of cooking, the use of multiple energy sources is relatively common in rural areas, and it is influenced by the diversity of energy supply, differences in functional requirements, and traditional factors.

Figure 6.12 shows the typical cooking situation in rural areas. The climate in northern China (such as Heilongjiang, Liaoning, Shanxi, and Gansu) is relatively cold, and residents are more inclined to use traditional cooking methods with coal or firewood, which can simultaneously meet the heating needs. Therefore, the electrification level in these regions is relatively low. Common kitchen and cooking equipment include large stoves, electric kettles, and gas stoves, while electrified cooking equipment such as range hoods, microwave ovens, and induction cookers are not common. However, the climate in southern China (such as Jiangsu and Hainan) is warmer, and residents tend to use simple and convenient electrified cooking equipment such as liquefied gas tanks (gas stoves), induction cookers, and rice cookers. Therefore, the electrification level in these areas is relatively high.

In terms of energy selection, electricity is gradually favored by residents in rural cooking, but in terms of total energy consumption, coal, and biomass are still the main energy materials for rural cooking. Under the premise of vigorously developing rooftop PV in rural areas in the future, rural cooking should also be vigorously electrified. Currently, the commonly used electric cookware can achieve full coverage and have a thermal efficiency of over 90%, far higher than the thermal efficiency of coal stoves and firewood stoves. At present, the commonly used electrified cooking equipment includes induction cookers, smart rice cookers, and integrated stoves. The induction cooker adopts the principle of electromagnetic induction and conducts the energy to the bottom of the pot through electromagnetic fields, with characteristics of high efficiency and energy conservation. Compared with traditional electric stoves, induction cookers have a faster heating speed, can realize more precise temperature control, and can meet the high-temperature requirements for cooking. The intelligent rice cooker has multiple cooking modes and reservation functions, by which, tasks such as cooking and stewing can be completed easily and quickly. As a new generation of kitchen appliances, integrated stoves are provided with the core functions of gas cooking appliances and range hoods. With the design of a side suction and bottom discharge structure of range hood, the stoves not only meet the thermal efficiency requirements of relevant national standards, but also further enhance the ability of range hoods, making the kitchen cleaner, more environmentally friendly, and healthier, better meeting the upgrading needs of consumers.

To promote electrification in energy consumption for rural cooking, it is necessary to comprehensively consider energy efficiency, power stability, sustainable development, and other factors. Electrification in cooking can be achieved by promoting efficient and energy-saving cooking equipment, combining the PEDF system, and

6.2 Electrification Solutions for Rural Energy Use in the Future

Fig. 6.12 Current situation of rural cooking in different regions of China

utilizing solar power generation. Among them, the flexibility in scheduling can ensure intelligent adjustment of the power consumption based on the power grid situation in rural areas, adapting to the load of the power grid. Besides, financial support and policy incentives can encourage rural residents to adopt electrified cooking equipment, improve the efficiency of energy utilization and the quality of life, achieving sustainable development.

6.2.4 Carbon Emissions Related to Agricultural Machinery and Evaluation on Electrification of Agricultural Machinery

In addition to energy consumption for living in rural areas, agricultural machinery plays a crucial role in energy consumption for agricultural production. According to the *China Rural Statistical Yearbook* from 2000 to 2020 and the provincial statistical yearbooks of 31 provinces (cities, and autonomous regions) in the Chinese mainland, the spatiotemporal distribution of carbon emissions from agricultural machinery in China from 2000 to 2020 was calculated.

Figure 6.13 shows the changes in carbon emissions from agricultural machinery and the proportion of carbon emissions from agricultural machinery to agricultural carbon emissions in China from 2000 to 2020. From the perspective of time, the national carbon emissions from agricultural machinery in 2000 were 8.29 $MtCO_{2e}$, which increased to 12.96 $MtCO_{2e}$ in 2015 (an increase of 56.4%) and slightly decreased to 10.91 $MtCO_{2e}$ in 2020 (a decrease of 15.9% since 2015). From the perspective of development trends, the change in carbon emissions from agricultural machinery shows an inverted U shape. The main reasons for this development trend include ① Measures or policies such as "two exemptions and three subsidies", "strengthening agricultural input and subsidies" and "mechanized agriculture" in Central Document No.1 from 2000 to 2015 promoted the vigorous development of agricultural mechanization, which led to an increase in carbon emissions from agricultural machinery; ② After 2015, the policy "promoting the green development of agriculture" and increasingly stringent emission standards in the No. 1 central document promoted the transformation of agricultural machinery to low-carbon, which led to a slight reduction in carbon emissions from agricultural machinery. Furthermore, from 2000 to 2020, the proportion of carbon emissions from agricultural machinery remained between 0.11 and 0.14. Among them, from 2004 to 2007, the proportion of carbon emissions from agricultural machinery increased significantly compared with other years, maintaining around 0.13. From this, it can be seen that the proportion of carbon emissions from agricultural machinery in the overall carbon emissions in agricultural production is relatively stable, and there is a possibility for emission reduction to a certain extent.

The cumulative carbon emissions from agricultural machinery vary greatly among different provinces. Figure 6.14 shows the cumulative carbon emissions from agricultural machinery in various regions over the past 20 years (2000–2020). From a spatial perspective, carbon emissions from agricultural machinery are mainly concentrated in the eastern plain areas, such as Hebei, Shandong, Henan, Jiangsu, Heilongjiang, and other main grain-producing areas, where carbon emissions from agricultural machinery have long been at a high level. As time went by, carbon emissions from agricultural machinery increased significantly in multiple provinces, such as Xinjiang, Shaanxi, Yunnan, and Guangdong; some provinces experienced a peak in carbon emissions from agricultural machinery in 2015 and then decreased to a certain

6.2 Electrification Solutions for Rural Energy Use in the Future 175

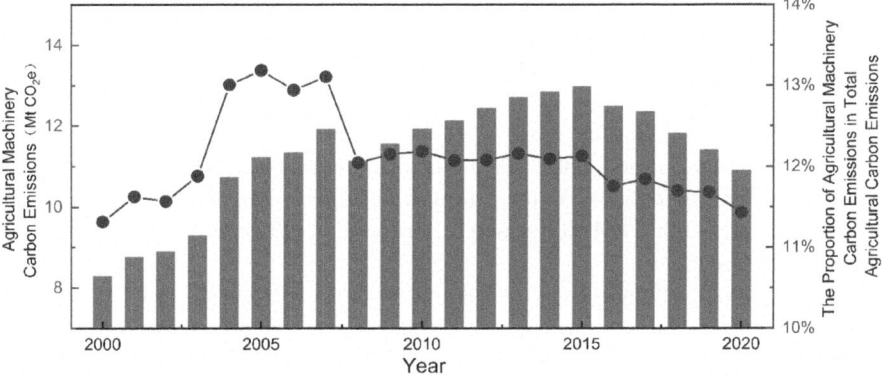

Fig. 6.13 Changes in carbon emissions from agricultural machinery in China from 2000 to 2020

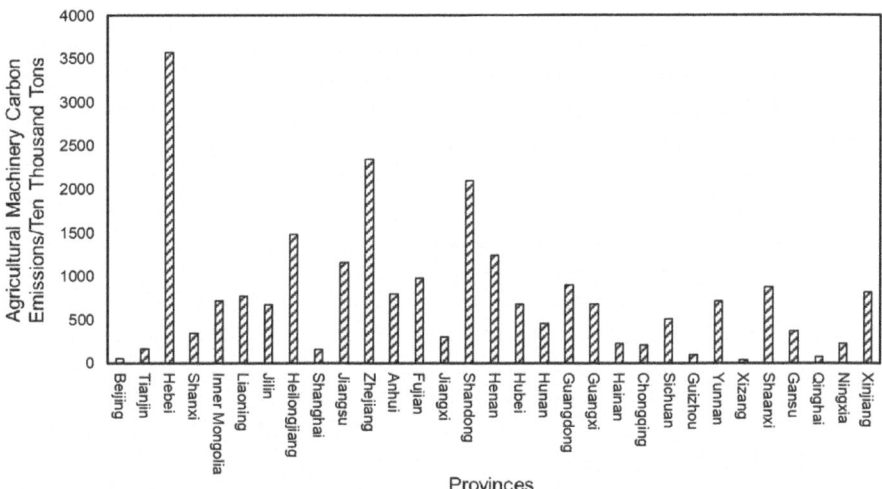

Fig. 6.14 Distribution of cumulative carbon emissions from agricultural machinery from 2000 to 2020

extent. From the perspective of the total quantity, there are five provinces with cumulative carbon emissions from agricultural machinery reaching 12 $MtCO_{2e}$, i.e., Hebei (35.73 $MtCO_{2e}$), Zhejiang (23.46 $MtCO_{2e}$), Shandong (20.97 $MtCO_{2e}$), Heilongjiang (14.78 $MtCO_{2e}$), and Henan (12.39 $MtCO_{2e}$)[1]. Among these five provinces, except for

---

[1] The 31 provinces in the Chinese mainland are put into three categories in terms of grain production function. The main grain producing areas include Hebei, Inner Mongolia, Liaoning, Jilin, Heilongjiang, Jiangsu, Anhui, Jiangxi, Shandong, Henan, Hubei, Hunan and Sichuan. Beijing, Tianjin, Shanghai, Zhejiang, Fujian, Guangdong and Hainan are believed to be main grain consumption areas. The areas with balance between grain production and consumption include Chongqing, Shanxi, Guangxi, Guizhou, Yunnan, Xizang, Gansu, Qinghai, Ningxia and Xinjiang.

Zhejiang, the other four provinces belong to the main grain-producing areas. The total cumulative carbon emissions from agricultural machinery in these five provinces are 107.35 MtCO$_{2e}$, accounting for 45.3% of the national total emissions. The cumulative carbon emissions from agricultural machinery in the main grain-producing areas are generally higher than those in other provinces. The total cumulative carbon emissions from agricultural machinery in the main grain-producing areas are 144.59 MtCO$_{2e}$, accounting for 60.9% of the national total emissions, 2.99 times that of grain consumption areas (48.34 MtCO$_{2e}$), and 3.26 times that of balanced grain production and consumption areas.

With the advancement of agricultural and rural electrification, the electrification of agricultural machinery will become a future development trend. The research and development of electric tractors started relatively late in China. In 1960, Harbin Songjiang Tractor Factory successfully trial-produced the first electric tractor in China, the Electric Bull 28, which was matched with a 28 kW AC induction motor, with a rated voltage of 1,100 V and no booster function. It is a cable-powered tractor [30].

Since the twenty-first century, with the research and development of electric tractors by relevant enterprises, higher education institutions, and research institutes, various technologies have developed rapidly. In 2012, YTO Group Corporation developed the Dongfanghong ET1400 Electric Tractor with alternating power supply from the power grid and batteries. It adopts a double motor structure, with a rated power of 5.5 kW for both the driving motor and the booster motor. It uses a lithium iron phosphate battery pack and can operate continuously for nearly one hour [31]. In 2018, the National Intelligent Agricultural Machinery and Equipment Collaborative Innovation Center created the concept prototype of an unmanned electric tractor—Super Tractor I, which passed the field operation test. However, the continuous operating time was only one hour, which made it difficult to meet the actual operating needs. In 2020, the Center released the first 5G + hydrogen fuel electric tractor (ET504-H) in China. The power system of ET504-H is mainly powered by hydrogen fuel cells and supplemented by lithium batteries, which can meet the needs of all the functions of traditional tractors. It is suitable for plains or hilly areas and can perform plowing, rotary tillage, seeding, and other operations according to the planned path in an unmanned mode. In 2021, Jiangsu Yueda Intelligent Agricultural Equipment Co., Ltd. successfully developed the Huanghai Jinma YL254ET Electric Tractor, which uses a lithium iron phosphate battery as the power source and adopts a double motor working mode, meeting the speed requirements for different machinery operations. In the same year, the project team of the National Key Research and Development Program "Intelligent Electric Tractor Development" (2016YFD0701000) developed two general-purpose electric tractors, SF350E and SF250E, with a continuous operating time of six hours, greatly improving the overall endurance and practicality of the equipment. In 2021, Shandong NOVAGRO Technology Co., Ltd. launched a lithium battery-powered unmanned tractor with intelligent functions, such as unmanned operation and collaborative operation. Honghu T150 Super Tractor, developed by the Intelligent Agricultural Machinery and Equipment Engineering Laboratory of

6.2 Electrification Solutions for Rural Energy Use in the Future

the Chinese Academy of Sciences in 2021, has a maximum power of 147 kW and supports rapid battery swap.

Since the first electric drive tractor emerged in the early twentieth century, significant progress has been made in related technologies with the improvement of power batteries, electric drive technology, overall and component control technology, as well as traction, power, and economic performance of electric tractors. However, the actual production and application scope of electric tractors is very limited, and there is still significant room for improvement in their performance, such as overall weight, endurance, and rated power.

Quantifying the emission reduction benefits of the electrification of agricultural machinery is conducive to accelerating the development and promotion of related technologies. A power prediction model for agricultural machinery was constructed based on five influencing factors, i.e., crop planting area, number of employed persons in the primary industry, number of persons engaged in rural and agricultural machinery industry, rural population, and total input in agricultural machinery. In consideration of the impact of planning policies, the Bass Diffusion Model was used to predict the penetration rate of electric agricultural machinery in the Chinese market. The greenhouse gas emission reduction benefits brought by the electrification of agricultural machinery were evaluated under multiple development scenarios (S1: low-speed development; S2: medium-speed development; S3: high-speed development).

According to Fig. 6.15a, as time goes by, the total power of agricultural machinery increases slowly and stabilizes eventually. The prediction results show that by 2025, the total power of agricultural machinery will be 1,136.87 GW (an increase of 7.4% since 2020). Besides, the results meet the master plan of "stabilizing the total power of agricultural machinery in China at around 1.1 billion kilowatts by 2025" in the "14th Five-year Plan for the Development of National Agricultural Mechanization" released by the Ministry of Agriculture and Rural Affairs. By 2030, the total power of agricultural machinery will be 1,188.5 GW (an increase of 4.8% since 2025), and by 2040, the total power of agricultural machinery will be 1,249.0 GW (an increase of 5.1% since 2030). According to Fig. 6.15b, under the market-driven scenario, the penetration rate of electric agricultural machinery in 2030 will be 3.85%, and it will be 10.3% in 2035 and 26.04% in 2040. The development trend under scenario S1 (low-speed development) is similar to that of the market-driven scenario, with slow growth in the initial stage and rapid growth in the later stage. Scenario S2 (medium-speed development) is approximately linear; under scenario S3 (high-speed development), the penetration rate of electric agricultural machinery has a rapid growth at first, followed by a slow growth. According to Fig. 6.15c, in combination with the predicted total power of agricultural machinery and the predicted penetration rate of electric agricultural machinery, the power for electrification of agricultural machinery was estimated. Under the market-driven scenario, the power for the electrification of agricultural machinery in 2030 is only 45.75 GW; under the scenario S1, S2, and S3, it is up to 112.90 GW, 285.12 GW, and 548.61 GW in 2030, respectively. From this, it can be seen that the penetration rates of electric agricultural machinery under the market-driven scenario and scenario S1 have a rapid growth momentum in the later stage, while the scenarios of proactive planning policies will promote the rapid growth of electric agricultural machinery in the initial stage; however, with the time going by, the development of electric agricultural machinery will tend to be stable finally. Besides, the scenarios of proactive planning policies can obtain more power for the electrification of agricultural machinery in the short term.

Figure 6.16 shows the carbon emission reduction benefits of the electrification of agricultural machinery under different scenarios. Under the existing standard scenario (DS) for emissions of traditional fuel, although there is a slight downward trend in carbon emissions from agricultural machinery, with the slow increase in power and comprehensive mechanization rate of agricultural machinery per unit area, carbon emissions from agricultural machinery will still show a slow growth trend, reaching 12.72 $MtCO_{2e}$ by 2030 (increase by 16.7% since 2020) and 13.09 $MtCO_{2e}$ by 2040 (increase by 2.9% since 2030). This indicates that it is necessary to implement the electrification transformation of agricultural machinery and the application of clean energy in the field of agricultural machinery. The electrification of agricultural machinery under the market-driven scenario will bring considerable emission reduction benefits. Under the market-driven scenario, carbon emissions from agricultural machinery will reach a peak of 12.38 $MtCO_{2e}$ in 2025, followed by a rapid decline. By 2030, an emission reduction of 0.51 $MtCO_{2e}$ will be achieved (compared with DS). The carbon emissions from agricultural machinery in the other three scenarios of planning policies will immediately decrease just after the implementation of the electrification of agricultural machinery, which means that planning policies will greatly shorten the carbon peaking and carbon neutrality process in the field of agricultural machinery. From this, it can be seen that the implementation of proactive planning policies will greatly promote the electrification process of agricultural machinery, bringing considerable carbon emission reduction benefits compared with the market-driven scenario.

The above is the evaluation of the emission reduction benefits of the electrification of agricultural machinery from a macro perspective, and the specific energy storage potential of agricultural machinery will be calculated subsequently. Based on the information obtained from relevant websites for domestic agricultural machinery, agricultural machinery is divided into two categories: transportation vehicles (road machinery) and production tools (non-road machinery). Among them, transportation vehicles mainly refer to crop handling vehicles, loaders, forklifts/stackers, small agricultural vehicles/dump trucks, and short-distance transportation vehicles that can operate on the road; production tools are further divided into power machinery, arable-land machinery, plant-protection machinery, harvesters, and processing machinery.

The machinery shown in bold and black in Fig. 6.17 is equipped with a power unit, while the machinery shown in gray generally does not have a power unit. Power machinery can be divided into walking tractors, crawler tractors, and wheel-hub tractors. Most agricultural machinery for arable land needs to be used in conjunction with power machinery. Machinery with power units includes small tillers and field management machinery. Among plant-protection machinery, there are seedling transplanters, transplanting machines, fertilizer applicators, and sprayers with power

6.2 Electrification Solutions for Rural Energy Use in the Future

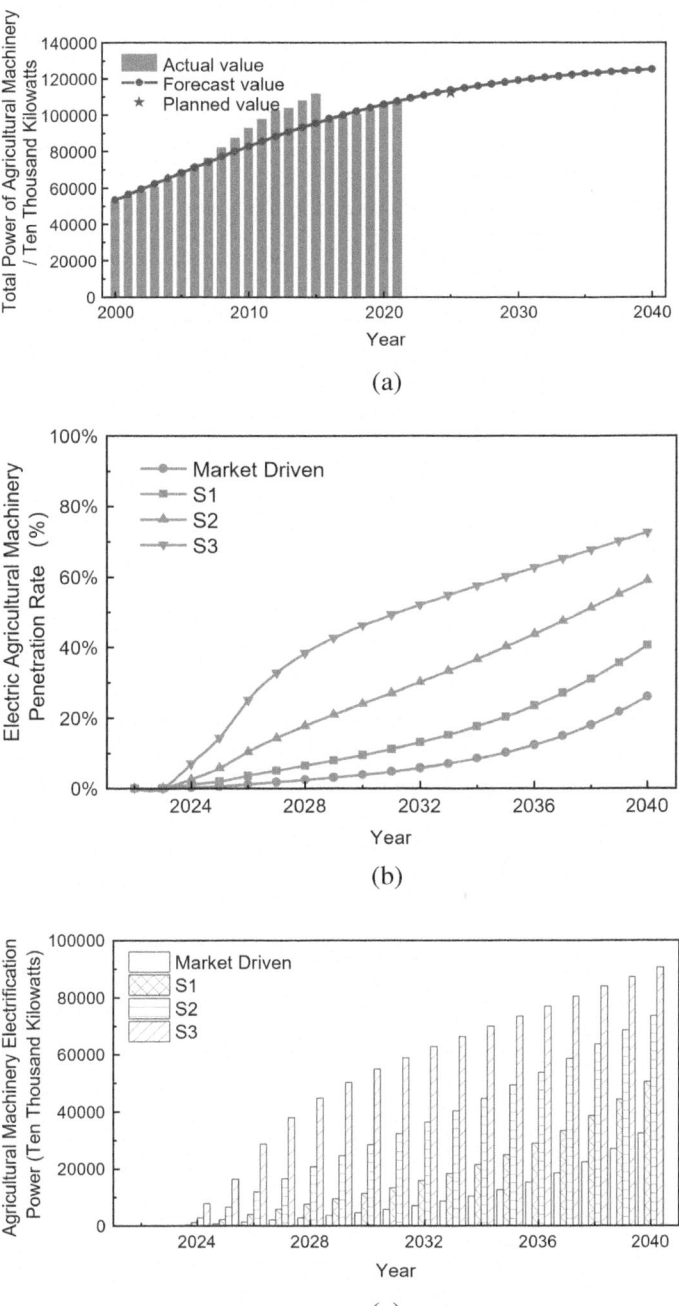

Fig. 6.15 Power prediction and electrification potential evaluation of agricultural machinery. **a** Prediction of total power of agricultural machinery; **b** Penetration rate of electric agricultural machinery; **c** Power for the electrification of agricultural machinery

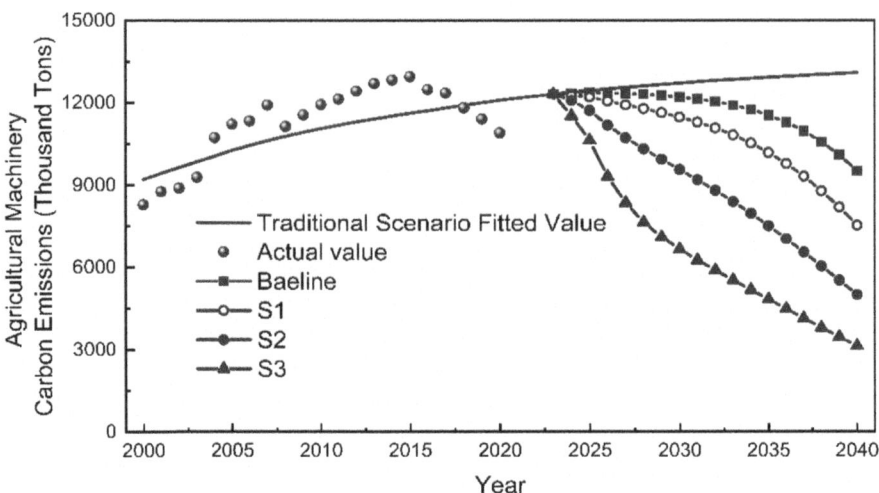

Fig. 6.16 Carbon emission reduction benefits of the electrification of agricultural machinery under different scenarios

units. Generally, seeders are also used together with power machinery. Harvesting machinery can be divided into wheat harvesters, rice harvesters, corn harvesters, and other economic crop harvesting machinery. There are threshers and hullers equipped with power units in processing machinery, and drying is generally carried out by sending them to the factory for centralized drying or air drying.

Based on the information obtained from relevant websites for agricultural machinery, the approximate power range of common agricultural machinery is summarized, as shown in Fig. 6.18. From the graph, it can be seen that the average power and power range of tractors and harvesters are relatively large. For tractors,

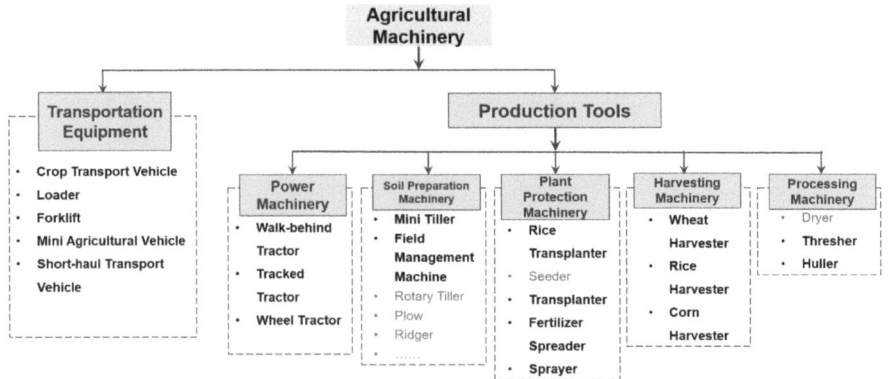

Fig. 6.17 Classification of agricultural machinery

6.2 Electrification Solutions for Rural Energy Use in the Future

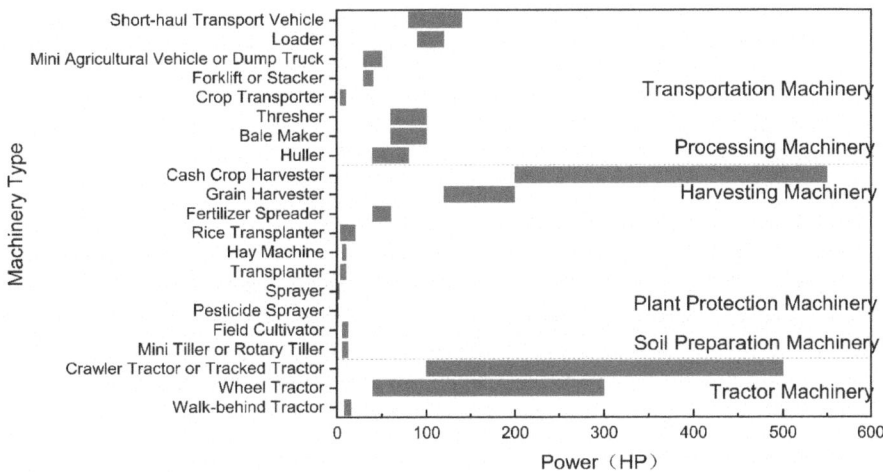

Fig. 6.18 Power of common agricultural machinery

the average power and power range of walking tractors is the smallest, approximately 8–15 HP, followed by the average power and power range of wheel-hub tractors, 40–300 HP, and the average power and power range of a crawler tractor is the largest, 100–500 HP. It can be seen that power machinery is often used in conjunction with other machinery, covering almost all power ranges. The overall power of the harvesting machinery is relatively high, with a maximum of 550 HP, and this is closely related to the complexity of the operating conditions and operational resistance of the machinery. The power range of arable-land machinery and plant-protection machinery is relatively low, and the power range of transportation machinery is related to the transportation distance to a certain extent.

The annual average operating time and fuel consumption of various agricultural machinery determined by research of the relevant records are shown in Fig. 6.19. This research covers 13 types of agricultural machinery including tractors, fertilizer spreaders, plowing machines, and harvesters. It can be seen from the figure that the level of operation of agricultural machinery varies with the type of machinery, the average annual operating time ranges from 217 h to 1,721 h, and the average annual fuel consumption of a single machinery ranges from 0.16 t to 10.50 t. The primary processing machinery of agricultural products and the field management machinery feature the maximum and minimum average annual operating time and fuel consumption, respectively. Tractors, planting and fertilizer machinery, tillers, and harvesters are often seen in agricultural production. Their average annual operating time is 588–669 h, 255–300 h, 348–427 h, and 450–607 h, and their average annual fuel consumption is 3.71–5.35 t, 1.30–1.98 t, 0.80–1.27 t, and 2.99–3.98 t, respectively. These four types of machinery have little variation in the level of operation. The post-harvest processing machinery, primary processing machinery, and earth working machinery used on farmland features the maximum range of variation in average annual operating time, and the primary processing machinery of agricultural products is the most variable in average annual oil consumption. This shows that the common crop planting machinery has a small range of variation in average annual operating time and average annual oil consumption; primary processing and other post-processing machinery, even similar machinery, somewhat vary in their levels of operation.

The energy storage potential of different types of agricultural machinery was estimated by summarizing the power and energy supply modes of such agricultural machinery and the battery parameters widely used in electric drive systems. As can be seen from Fig. 6.20a, tractors have the largest range (11–589 kWh) of energy storage potential. This is ascribed to its combined characteristics as a power unit. According to the survey, the current enterprise research and development (R&D) of electric

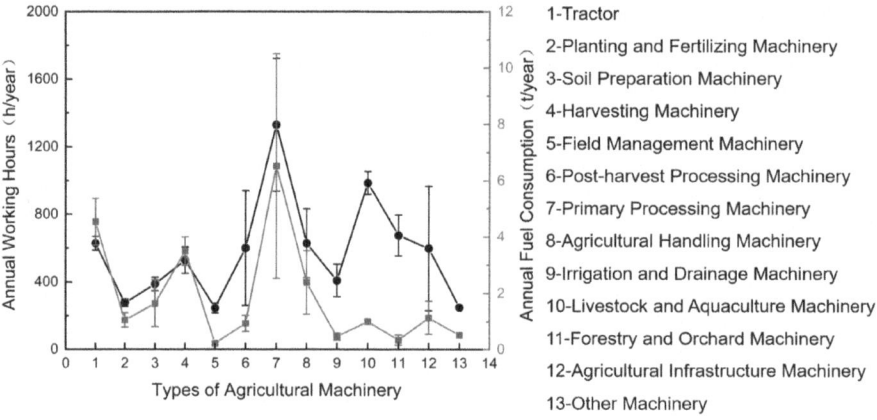

Fig. 6.19 Levels of operation of different types of agricultural machinery

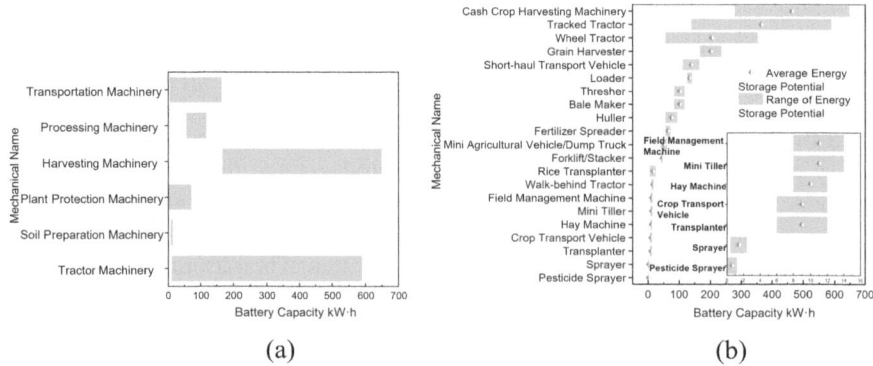

Fig. 6.20 Estimation of energy storage potential of agricultural machinery. **a** Agricultural machinery; **b** Specific equipment

6.2 Electrification Solutions for Rural Energy Use in the Future

tractors is mainly focused on machinery with a power of about 50 HP; few R&D efforts are made on high-horsepower electric tractors, given the difficulty. The energy storage of harvesting machinery ranges from 168 to 648 kWh. However, there is no relevant research available on electric harvesting machinery. The compatible portfolio of plant protection machinery and tillers exhibits great energy storage potential, 0.04–71 kWh and 8–14 kWh, respectively. The energy storage potential of transport machinery is 6–165 kWh, and the electrification of such equipment is most similar to the development of electric vehicles. As can be seen from Fig. 6.20b, the top five types of agricultural machinery in terms of average energy storage potential are economic crop harvesting machinery, crawler tractors, hub tractors, grain crop harvesters, and short-distance vehicles. The first three types of agricultural machinery boast relatively large ranges of energy storage potential, and the remaining agricultural machinery is relatively small in the energy storage potential range, averaging within 100 kWh. The last seven types of agricultural machinery are pesticide sprayers, sprayers, transplanters, crop haulers, hay mowers, mini-tillers, and field management machinery. Their energy storage potential is within 10 kWh. They are commonly seen in peasant households, and can basically meet the energy storage scheduling requirements of the new household power system.

Where the rural areas have built a new energy system based on rooftop PV in the future, the energy storage is provided by various types of electric equipment for the system. First of all, under the scenario of full rural electrification in the future, the energy storage resources in rural areas fall into three categories: production, transportation, and large village-level machinery. Electric production equipment mainly refers to agricultural machinery commonly seen in peasant households, such as hay mowers, sprayers, pesticide sprayers, and mini-tillers. It is estimated that such equipment can provide at least 9 kWh of energy storage potential. Electric transportation equipment mainly refers to the common means of transportation in peasant households, such as electric bicycles, electric tricycles, and low-speed electric vehicles. The electric tricycles even function as agricultural transporters. It is estimated that such equipment can provide at least 9.5 kWh of energy storage potential. In terms of large village-level machinery, electric equipment mainly refers to the large agricultural equipment at the village level or shared by more than one village, e.g., farm transporters, tractors, and forklifts. In summary, the production, transportation, and large village-level machinery can provide at least 20 kWh of energy storage potential for the new rural energy system at the household level, which can meet the energy storage scheduling requirements of peasant households.

The CPC Central Committee issued Central Document No.1 on agriculture, rural areas, and farmers for five consecutive years from 1982 to 1986, making specific plans for rural reform and agricultural development. The Central Document No. 1 on the theme of agriculture, rural areas, and farmers was issued for 19 consecutive years from 2004 to 2023, highlighting the "top priority" of the issues relating to agriculture, rural areas, and farmers in China's socialist modernization. The Document pointed out that through the orderly transfer of rural cultivated land, the dispersed land of farmers was converted to large-scale management, which realized the orderly and moderate intensification of agricultural land, and promoted the agricultural transformation and upgrading such as the large-scale operation of agricultural machinery and construction of farmland infrastructure. With the decrease in rural population, agricultural production inevitably becomes intensive and automated. In this context, the types of electric agricultural machinery will increase significantly, and household-based electric agricultural machinery and equipment will orderly and fully interact with the large-scale electric agricultural machinery and equipment through battery cabinets and assembled batteries.

6.2.5 Suggestions for Increasing the Level of Electrification of Energy Use in Agriculture and Rural Areas

There is a certain gap between the level of electrification of household energy consumption in rural areas and that in urban areas, to the extent that the R&D and application of electrified machinery for agricultural production are in their infancy. To achieve the dual carbon goals and promote the transition to cleaner energy use in rural areas, full electrification is the main path of future rural energy development. To this end, we should work on the following aspects to improve the overall level of electrification of energy use in agriculture and rural areas.

First, vigorously develop distributed clean energy and establish a rural power system with rooftop PV as the core. More efforts will be made to advance the distributed clean energy grid connection in rural areas, and ensure the timely completion of power grid upgrading projects, in order to achieve the full consumption of clean energy. In areas where conditions permit, we will build village and town demonstration plots for comprehensive rural energy, and promote the development of circular, ecological, and comprehensive energy that combines smart energy with the green ecosystem.

Second, boost the electrification of agricultural machinery. Given the carbon emissions of agricultural machinery, the level of agricultural development, and the characteristics of regional demand, it is essential to formulate differentiated policies for the electrification of agricultural machinery. Moreover, to increase the popularity and market size of electric agricultural machinery, the government is advised to maintain the continuity and stability of incentive policies, such as tax exemption for the purchase of electric agricultural machinery, and to provide financial and technical support. Meanwhile, agricultural machinery manufacturers should be encouraged to transform and upgrade electric agricultural machinery. With national financial and policy support, enterprises may increase R&D investment and plant capacity, improve product quality and performance, and reduce the production cost and selling price of electric agricultural machinery, in order to improve farmers' recognition and willingness to use electric agricultural machinery. This can escort the continuous promotion and use of electric agricultural machinery. Further, we can promote the upgrading of agricultural electrification, the electrification of rural family farms, and modern agriculture, in an effort to provide power support for new rural forms of business like e-commerce, storage and preservation, and cold chain logistics.

Third, vigorously stimulate the potential of household energy consumption for rural residents, and improve the level of electrification in rural life. We will explore low-carbon and even zero-carbon rural residential building pilots, promote the application of energy-efficient equipment, and popularize green energy use in rural areas. Efforts are made to promote the transformation of traditional energy use habits in rural households by popularizing the application of household appliances (electric steamers, induction cookers, electric ovens, etc.), household electrical equipment (e.g., conditioners and electric heating products), and cleaning and sanitary equipment (air source water heaters, etc.). We will strive to facilitate rural electric mobility and accelerate the construction of public charging facilities in rural areas to support green transportation. More rural battery electric vehicles are encouraged to be used in agricultural vehicles and other fields. Electric boats and other means of transport will be gradually introduced in rural areas where conditions permit. In addition, DC charging facilities can be built in key areas such as rural public undertakings, transport hubs, trade logistics bases, and tourist attractions to meet the requirements of rural electric mobility.

6.3 Rural PEDF Technical Points and Presentation of Village-Level Demonstration Cases

Unlike traditional rural rooftop PV projects, the rural PEDF project is not simply designed to develop rooftop PV; instead, it is ambitious to boost the full electrification of rural energy use by urging the local consumption of PV power in rural areas. Besides, it's endeavoring to achieve flexible interaction with rural transformers, and break the restrictions of rural transformers on the PV promotion in rural areas. Allowing for its volatility and randomness, the PV power suffers a serious mismatch to rural loads. To secure the stability of PV-micro-grid voltage in rural areas and meet the real-time power supply requirements, two technical challenges must be addressed: ① accurate and rapid control technology; ② low-cost energy storage solutions. Meantime, the rural PEDF system needs to solve the problems: equipment selection combined with optimal calculation, the rational economy-based layout of the system, DC transformation scheme of equipment, operation strategy in response to the large grid dispatching, and PV heating solutions for northern rural areas. The PEDF demonstration projects must realize the basic principle of "self-use prioritized, and orderly feed-in of surplus electricity", and further meet two quantitative indicators: ① The PV self-satisfaction rate is more than 80%, that is, the vast majority of the electrical load in rural households and villages is supplied by PV power, thus facilitating the local PV consumption; ② the installed PV capacity is more than 1.5 times the peak power of the system and transformer, so as to break the limitation of transformer capacity on the promotion of rural PV systems. Let's take a village in

Shandong Province as an example to illustrate how to resolve the mismatch between supply and demand faced by PV consumption through technological innovation and reasonable scheme design, as well as the transformation of RRB loads and energy storage resources. This section presents the quantitative evaluation indicators, which can be used as a reference for evaluating the feasibility of rural PEDF projects.

6.3.1 Basic Information

Village A is located in Kenli District, west of Dongying City, Shandong Province. The original inhabitants of the village have been relocated to a rural community about 1 km to the northeast, and the RRBs in the village are now used as homestay inns. There are two reasons for choosing Village A as the demonstration site of rural PEDF projects: ① With the aggravation of rural population aging and hollowing-out in China, the hollowed village consolidation is a way of rural future development. To some extent, Village A represents the future rural scene of our country; ② Village A has typical rural load characteristics, ranging from cooking, heating, and electrical loads to vegetable greenhouses, a grain drying plant, and other agricultural production loads around the village. The installed PV capacity of the PEDF demonstration project (Phase I) in Village A is 300 kW, and a total of 26 RRBs are transformed to form a transformer coverage area.

The following will showcase the design schemes of the PEDF power system in Village A from local to global: the RRB heating solution (heating load), the RRB microgrid design scheme (electrical load), the village-level microgrid design scheme, and the solution for agricultural production load and energy storage.

6.3.2 RRB Heating Solution

In the northern rural areas, heating energy consumption accounts for more than 70% of the total energy consumption of RRBs. What's more, farmers rely too much on fossil energy for heating (coal-fired furnaces, heated brick beds, etc.), so heating in northern China releases a lot of pollutants and CO_2. Therefore, the heating of RRB is an inevitable challenge in the new rural power system. The biggest difficulty of PV power heating lies in the mismatch between supply and demand. Specifically, PV power is climaxed at the maximum illumination intensity, when the RRB requires the lowest heating load; there is zero PV power at night, but the demand for heating RRB is still there. The PV heating system must respond to the challenge of how to avoid overheating at noon and underheating at night. Obviously, heating products such as PV-driven graphene sheets, which are currently popular in the market, cannot solve the mismatch between PV power and heating load. Allowing for the relatively low income of farmers, only low-cost heating can be popularized in rural areas. The economy is another challenge facing the PV heating system of RRB.

Allowing for the scattered rooms in rural areas, the RRB rooms can fall into "full-time heating" and "partial-time heating" by usage pattern. "Full-time heating" includes rooms that are heated all day long, e.g., bedrooms. The heating load of such rooms is relatively stationary. By means of energy storage, transferring excess energy during the daytime to the night is an effective way to balance the supply and demand between PV power and heating load. Nevertheless, batteries are too expensive for farmers to afford. The prices and space requirements of water heaters and buried pipes are also more than farmers can afford. Therefore, the Tsinghua University team used cheap red bricks as heat storage materials in Ruicheng County, Shanxi Province, and connected carbon fiber heating wires directly with the PV power system as the heating terminal in the RRB heating demonstration. Carbon fiber heating wires are evenly laid on the inside of the northern exterior wall with insulation, and then the carbon fiber is covered with a red brick wall. In this way, the carbon fiber converts PV power into heat, which heats the red brick wall by heat conduction, and the red brick wall heats the room by convection and radiation. Thanks to the thermal inertia, red bricks can release heat into the room at night to maintain the indoor temperature relatively stable. The control device was abandoned for cost reasons. Therefore, an accurate prediction model is required to determine the design parameters to avoid safety accidents as a result of overheating of heating wires or the poor heating effect resulting from insufficient capacity allocation. A detailed description and results of the solution are given in 6.3.

Red bricks are a better way to store heat for new buildings. With a similar principle, the PV + electric storage heater scheme applies to existing buildings. The popular electric storage heaters in the market suffer the following disadvantages: ① improper temperature control mode. The existing electric storage heater uses the temperature on the surface of the ceramic tube to control the ON/OFF of the electric heater, which wastes a lot of heat storage potential. If connected to the PV system, it will result in a lot of "solar curtailment". ② Single operating mode. The electric heater can only rely on natural convection generated by the heating of the ceramic tube surface to heat the room. At a low surface temperature, the ceramic tube or accumulator tube can hardly provide enough heat for the room. Therefore, the Tsinghua University team designed a new PV-driven electric storage heater. First, the temperature control strategy is changed by installing the temperature probe on the outer surface of the heat storage bricks. This way allows the regulation of operating conditions by detecting the temperature of heat storage bricks, maximizing the heat storage potential, and avoiding "solar curtailment". Second, the intelligent convection device is installed at the bottom of the electric heater to adjust the operating state by sensing the outlet air temperature of the equipment. When the heat storage brick is cold, the heat transfer is increased by mechanical convection to ensure the heating requirements of the room.

The "partial-time heating" rooms are known for unsteady heating loads, and the heat storage may reduce the flexibility of heating equipment. In addition, the direct electric heating equipment features low energy conversion efficiency, and the energy efficiency of PV systems can be improved significantly by PV direct-drive air source heat pumps. For such rooms of partial-time heating with high-efficiency heating devices, the low ambient temperature air-to-air heat pump is an effective way to address the demand for partial-time heating. However, the portfolio of air-to-air heat pumps and PV power requires an accurate and fast control strategy to ensure the stable voltage of the system, otherwise the fluctuating voltage of PV power can hardly meet the operation requirements of the air-to-air heat pump. For DC systems, the voltage signal is the most concise and direct information that indicates the running status of equipment. Therefore, the Tsinghua University team adopts the method based on the busbar voltage signal to achieve voltage stability on the supply side, and in the meantime, the power supplied to the air-to-air heat pump is balanced with the demand power of the air-to-air heat pump in real-time, so as to ensure the stable operation of the air-to-air heat pump.

6.3.3 RRB Cooking Solution

Characterized by high power and short duration, kitchen appliances pose great challenges to the stable operation of the whole indoor microgrid system. Therefore, the DC transformation and moderate flexible response regulation of kitchen appliances are an important part of achieving efficient energy utilization and optimizing the PEDF system. Induction cookers are common high-power appliances in kitchens and are traditionally designed to be AC-powered. To enable DC supply, it is necessary to replace the AC-DC rectifier module with the DC-DC power conversion module, so that it can accept DC input and provide high-frequency current suitable for induction heating. Meantime, the control circuit is redesigned to accommodate the DC input and ensure the precise control of output power to suit different cooking requirements. However, direct current may cause a larger spark, and the safety characteristics of induction cookers, such as short-circuit protection and overvoltage protection, are re-evaluated and enhanced. Rice cookers usually include a heating element and a microprocessor controller. The heating element needs to be replaced or modified during DC transformation so that it can be powered directly from DC. Meanwhile, the controller should be updated to accommodate DC inputs and ensure that it can adjust the heating power according to the DC power supply changes to maintain cooking efficiency and food quality. The DC transformation of microwave ovens is relatively complicated because they use a high-voltage AC power supply to drive the magnetron. First, it is essential to design an efficient DC-DC converter that can provide the magnetron with the required high-voltage DC supply. Second, a new control system must be developed to manage the DC power supply and ensure the normal operation and safety of microwave ovens. The DC transformation of electric kettles is simple by replacing the heating element to adapt to the DC power supply. Based on the above DC transformation, an integrated kitchen energy management system is required to monitor and regulate the energy consumption of all kitchen appliances. The system should be able to monitor the DC busbar voltage and the energy consumption of kitchen appliances in real-time. When fully powered, it optimizes the operating efficiency of equipment, and reduces energy consumption or suspends non-key equipment in case of power shortage. Moreover, it is necessary to ensure the safety of all equipment when operating under a DC power supply.

6.3.4 Household Microgrid Solution

Determining the system topology is the first step to building the household microgrid of a new rural power system, and also the basis of the whole scheme design. The topology of the new rural PEDF power system can be roughly divided into the following three types (by the frame structure only, not by details such as voltage level and the location of electrical storage devices), as shown in Fig. 6.21.

Different topologies result in different power generation efficiencies, transmission losses, equipment capacity, control methods, and investment costs. First of all, the selection of equipment capacity will vary with the location and connection power of the converter, thus giving rise to changes in system costs. For example, the busbar converter directly connected to PV in Scheme 2 needs to determine the capacity based on the installed PV capacity, while the busbar converter in Scheme 1 and Scheme 3 needs to consider the household loads only. Second, the power generation benefits, and conversion transmission losses vary with the scheme. The PV power in Scheme 1 and Scheme 3 requires only a single converter, while Scheme 2 requires two converters, resulting in high power generation loss. Besides, the low access voltage of PV power in Scheme 2 will create more line current, thereby increasing the line loss. Finally, the choice of control strategy varies from scheme to scheme. A significant advantage of DC systems is that decentralized busbar voltage sag control can be used to ensure stable system operation. This can reduce controller investment and maintenance costs. In Scheme 1, the converter is not connected directly to the busbar and cannot respond quickly to voltage changes, making the voltage sag control not possible. Instead, additional cost is required for an embedded controller. The installed PV capacity of 10 kW is taken as an example to compare the three schemes, as shown in Table 6.1.

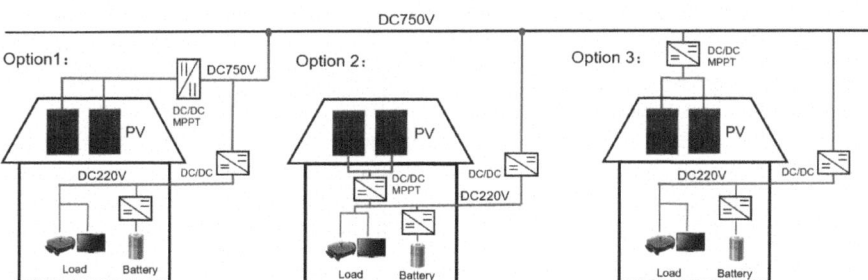

Fig. 6.21 Three different power system topologies

Table 6.1 Comparison of household power topology schemes

Item	Scheme 1	Scheme 2	Scheme 3
Power generation efficiency	95%	95%95%	95%
Transmission loss	Low	High	Low
PV converter capacity	10kw	10kw	10kw
Busbar converter capacity	3 kW	10kw	3 kW
Storage converter capacity	Identical		
Control method	MPC control	Busbar voltage sag control	
Disadvantages	The investment and maintenance of the controller entail additional costs	Converter costs and losses are relatively high	It does not meet the "self-use prioritized" principle

Compared with the 10 kW installed PV capacity shown in Table 6.1, the three schemes have their own advantages and disadvantages. Therefore, the actual project should be considered according to the local electricity price, equipment cost, grid-connected policy, and flexible targets to determine the most cost-effective scheme. Scheme 2 is preferred for power systems that follow the principle of "self-use prioritized, and orderly feed-in of surplus electricity". Despite the low energy production, large loss, and relatively high capacity and cost of the converter equipment, the total system cost is reduced, benefiting from the simplified control modes and the low overall control cost. As the power electronic equipment technology advances in the future, the cost of converters will be reduced, thus narrowing the cost gap caused by capacity differences. With the increase of energy storage equipment in RRB, the power flowing through bus converters will decrease, and so will the requirements for equipment capacity. In view of the low electricity prices for rural residents, the system conversion and transmission losses cause relatively limited economic losses.

The DC transformation on the indoor load side is beneficial to further improving the conversion efficiency of the system, and is more suitable for sensing the signals of busbar voltage, so as to adjust the flexible response in real-time. First, we must carry out fine classification of household electrical equipment, and design a DC transformation scheme for each type of equipment. Second, an intelligent energy management system shall be built to monitor and regulate the energy consumption of various loads, and to achieve the flexible response based on DC busbar voltage. Thanks to the inherent DC working nature, lighting facilities, such as LED luminaires, are simple in transformation by directly connecting to the DC busbar. In this process, the smart lighting system can adjust the luminance of luminaires according to the real-time voltage level, so as to perform dynamic response to voltage fluctuations and optimize energy consumption. The AC/DC converters inside entertainment electronics, say TV and computers, need to be replaced with a DC input module, or are powered by an external DC power adapter. Allowing for user satisfaction and the fact that devices with high user frequency cannot be adjusted during certain periods, such devices shall be connected to the intelligent power management system for the sake of stability. In this way, the power supply and scheduling are prioritized, and the power consumption of non-key equipment is first reduced when the voltage drops. Large household appliances (refrigerators, washing machines, etc.) are usually driven by AC motors; AC motors need to be transformed into DC motors, or DC power supply modules are connected to such household appliances. Due to the unadjustable power waveform of equipment such as washing machines, the intelligent energy management system can only adjust the operating time of washing machines according to voltage changes, thus responding actively to flexible scheduling without affecting the user experience.

For home charging loads, such as mobile phones, tablet computers, electric bicycles, and electric vehicles, the interface of mobile equipment chargers (mobile phones, tablet computers, etc.) has a DC version available, and they can be charged directly by the DC power supply. Therefore, the DC transformation only needs to replace the traditional AC adapter with a DC adapter. Charging points for electric vehicles typically receive AC and convert it internally to DC to charge the vehicle battery. Hence, the DC transformation requires the provision of special DC charging points, which can be connected directly to the DC busbar and are capable of providing direct current for electric vehicles.

An integrated smart home energy management system will be an indispensable hub throughout the DC transformation process. This system needs to monitor the energy consumption within the home and the DC busbar voltage level in real-time, and can intelligently deploy loads, so that the stable system operation and the optimal energy distribution can be maintained under different voltage conditions. Such system design and intelligent management can improve energy efficiency, meet the daily electricity demand of users, and provide an effective buffer against fluctuations in energy supply.

Here is an example of a sunny day to analyze the operation process of an indoor microgrid system:

(1) Before 8:00 in the morning, there is no PV power, as shown in Fig. 6.22. The electrical load steps down the indoor busbar voltage. When the indoor busbar voltage U_n is lower than the battery discharge set voltage U_{b2}, the indoor battery discharges to keep the voltage stable and provide power for the load.

(2) From 8:00 to 12:00, the PV power generation meets the load demand, with low light, as shown in Fig. 6.23. The indoor busbar voltage is lower than the PV converter set voltage U_m, and the PV system operates in the maximum power point tracking (MPPT) state. If the indoor busbar voltage U_n is higher than the battery charge voltage U_{b1}, the indoor battery is charged. If the indoor busbar voltage U_n is higher than the busbar converter set voltage U_{c1}, it supplies power to the village-level busbar.

(3) From 12:00 to 17:00, the PV power generation meets the load demand, with high light, as shown in Fig. 6.24. The indoor busbar voltage U_n is higher than the

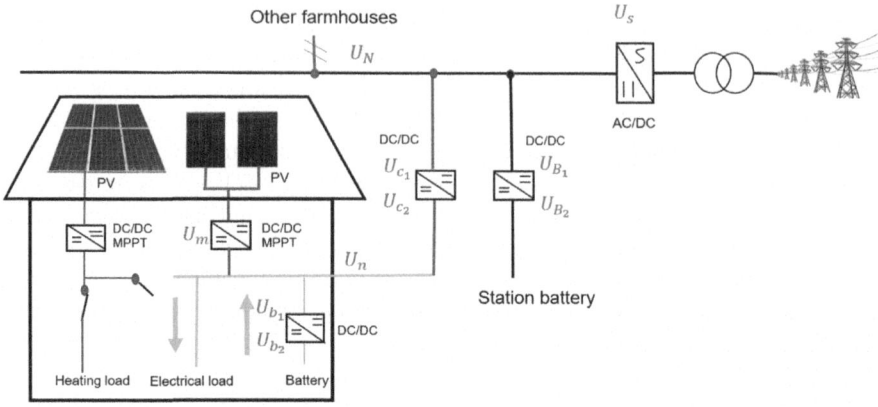

Fig. 6.22 No PV power scenario

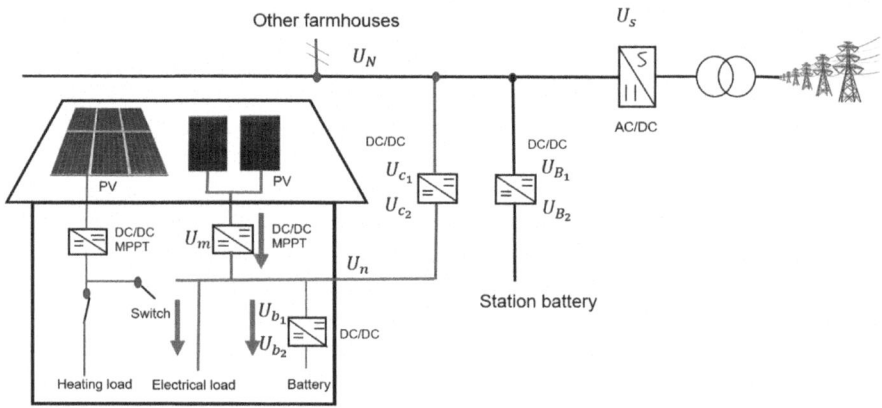

Fig. 6.23 Low light scenario

PV converter set voltage U_{C1}, and the PV system operates at constant voltage. After 13:00, power is supplied to the village-level busbar. The village-level busbar voltage U_n is higher than the PCS set voltage U_s, and the transformer coverage area supplies power to the grid.

(4) From 18:00 to 23:00, it is the same as the scenario (1). The change-over switch of heating PV panels is always connected with the heating equipment to form a stand-alone system only for heating. In the non-heating season, the change-over switch is connected to the indoor busbar, solving the indoor electricity demand jointly with the electrical PV panels.

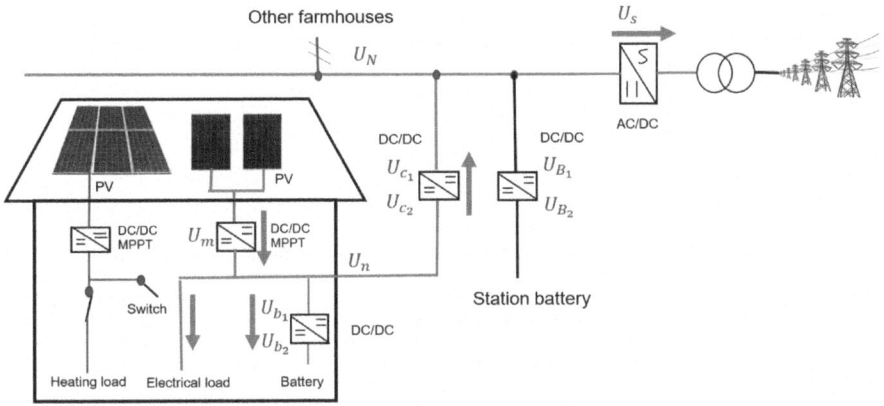

Fig. 6.24 High light scenario

6.3.5 Village-Level Microgrid Design Scheme

The indoor microgrid in the RRB supplies the excess PV power to the village-level busbar. The village-level busbar is also connected with the public space and equipment, such as vegetable greenhouses, streetlights, and the Party-mass activity center. In this way, the rooftop PV power of RRB can be further consumed by the electrical load in the village. The energy storage in the RRB can secure the electricity demand of farmers at night, and the village-level microgrid has a large-capacity energy storage system available. With the process of agricultural electrification, electric agricultural machinery will become a "free" village-level energy storage resource (see 6.3.6 for specific schemes). Finally, the village-level busbar is connected to the PCS system to dispatch excess power in response to the grid into the large power grid. This section introduces the operation process of village-level microgrids, combined with specific scenario analysis.

During clear daylight hours, the installed PV capacity is far greater than the electrical load of RRB. Therefore, after the PV power meets its own electricity demand and indoor battery charging demand, the excess power is supplied to the village-level busbar, as shown in Fig. 6.25. After the village-level busbar power meets the public load demand in the village, the excess power is regulated by the village-level energy storage system and then supplied to the grid according to the dispatching of the large power grid. According to the rules of busbar voltage signal control, power will be supplied to the large power grid when the village-level busbar voltage exceeds the PCS set voltage. When the large power grid issues an instruction of no or reduced power supply to the rural microgrid, the PCS will receive the instruction to raise the set voltage. Meantime, the village-level energy storage system increases the charge power to step down the village-level busbar voltage and reduce the power of the village-level microgrid to the large power grid. When the large power grid issues an instruction to increase the power of the rural microgrid, the PCS will receive

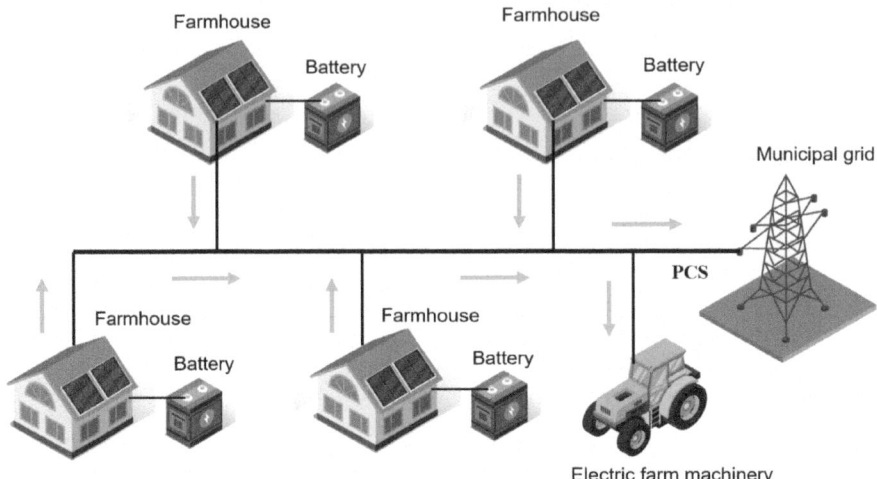

Fig. 6.25 Operating state of village-level microgrid during the clear day

the instruction to lower the set voltage. Meantime, the village-level energy storage system reduces the charge power and even discharges to the village-level busbar to step up the village-level busbar voltage and increase the power of the village-level microgrid to the large power grid.

On clear nights, the electricity demand of RRB is satisfied by indoor batteries. If the next day is still sunny or cloudy, the village-level energy storage system will supply the stored electricity to the large power grid to gain revenue from electricity sales, as shown in Fig. 6.26. If the next day is rainy, the village-level energy storage system will conserve power to meet the energy demand for villagers on the following rainy days, as shown in Fig. 6.27.

For village-level microgrids, different energy storage configurations will lead to different operation strategies. The energy storage functions to meet the demand for electricity in rural areas, and to provide flexibility for the operation of microgrids. As a result, different energy storage capacity configurations result in diverse grid interaction strategies, transformer capacity requirements, grid dispatching flexibility, and initial investment costs. In this sense, the energy storage capacity of the system has to be optimized according to the system running target parameters and the economy.

Figure 6.28 compares the operating conditions of Village A for 24 h on a sunny day with different energy storage capacities. The energy storage capacities include village-level energy storage and indoor energy storage. As can be seen, without the energy storage availability, the village-level microgrid has to draw power from the large power grid at night, and requires a transformer capacity of 340 kVA (80% to meet the peak power interacting with the grid). With the 110 kWh energy storage capacity, the energy demand of peasant households at night can be met by energy storage; there is no need to draw power from the large power grid, thus reducing the dependence on the large power grid. The 240 kWh energy storage capacity can

6.3 Rural PEDF Technical Points and Presentation of Village-Level ...

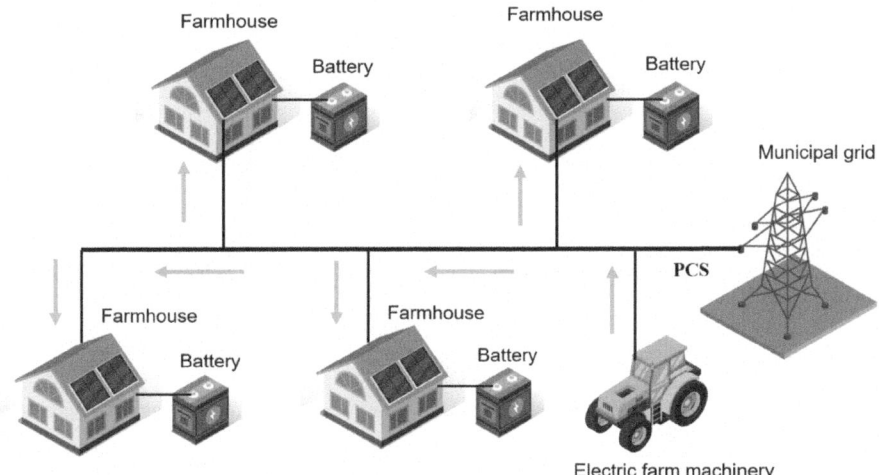

Fig. 6.26 Operating state of village-level microgrid on consecutive clear nights

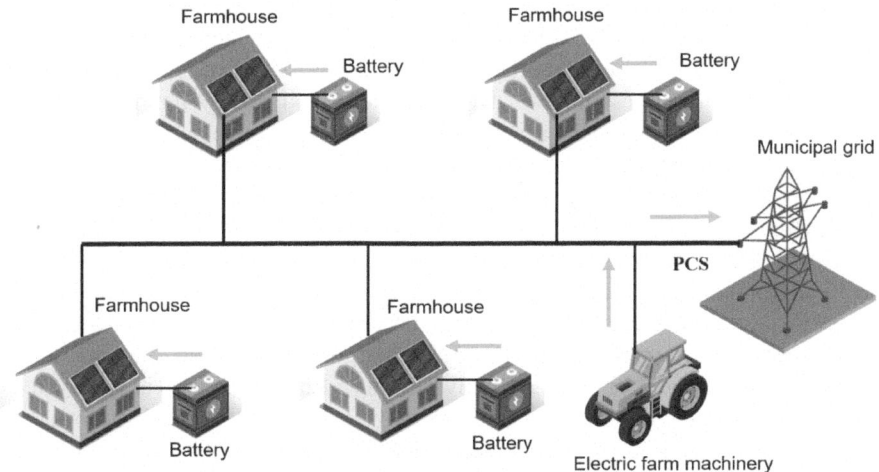

Fig. 6.27 Operating state of village-level microgrid on rainy days

meet the electricity demand of peasant households, and significantly reduce the peak power interacting with the grid, thus reducing the expansion pressure of village-level transformers by about 100 kVA. In the case of 360 kWh energy storage capacity, the village-level microgrid can even supply power to the large power grid at night to gain additional revenue from electricity sales. Moreover, the ratio of installed PV capacity to transformer capacity increases from 1.1 without energy storage to 1.7, indicating that the village-level microgrid breaks the limitation of transformer capacity for installed PV capacity.

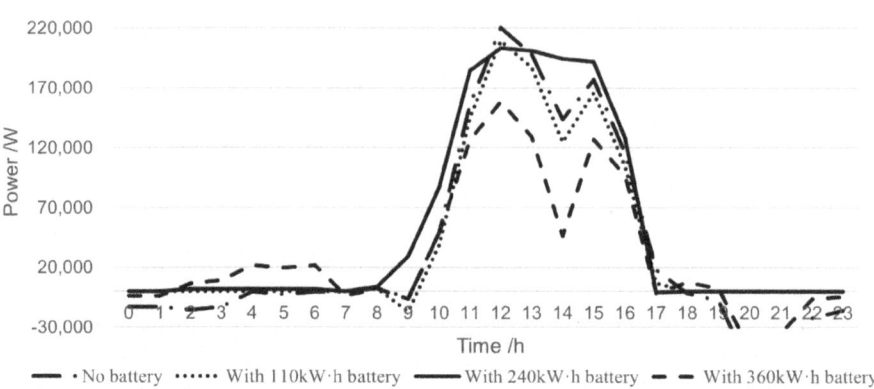

Fig. 6.28 Comparison of village-level microgrid operation under different energy storage capacities

6.3.6 Production Load and Energy Storage Solutions

With the continuous advancement of the electrification of agricultural production equipment, the synergy of the PEDF system and electric agricultural machinery will greatly improve agricultural production efficiency and renewable energy utilization. The rural microgrid supplies daily electricity and energy required for agricultural production, and even carries out intelligent power interaction with electric agricultural machinery batteries. The core of this strategy lies in that the battery pack of electric agricultural machinery acts as its power source, and as a dynamic energy storage unit in the microgrid. That is, the rural rooftop PV power is used to meet the battery charging requirements of agricultural machinery, and agricultural machinery batteries provide energy storage services for the microgrid. In this way, the electric agricultural machinery becomes the battery of the village-level microgrid, and the village-level microgrid becomes the charging point of electric agricultural machinery. This model requires a highly automated battery swap cabinet system, which enables fast and efficient swapping of battery packs, greatly improves the efficiency of agricultural operation, and optimizes the energy distribution on the microgrid. The village-level energy storage system based on standardized batteries for agricultural machinery includes the following parts:

(1) Intelligent control system: The system is essentially an advanced control system that coordinates the battery pack charging, swapping, and grid energy management for electric agricultural machinery. The system can monitor the PV power generation, grid load, battery status, and agricultural operation requirements in real-time.

(2) Standardized battery module: The electric agricultural machinery uses unified standard battery modules, which are designed to be quick plug-ins for quick swapping in the battery swap cabinet. This modular design allows the battery to be used across different types of agricultural machinery and battery swap cabinets.

6.3 Rural PEDF Technical Points and Presentation of Village-Level … 197

Fig. 6.29 Standardized battery swap cabinet

(3) Battery swap cabinet: As shown in Fig. 6.29, a battery swap cabinet is a physical interface that connects the electric agricultural machinery to the microgrid. It can automatically swap, charge, and maintain battery packs. The battery swap cabinet is internally equipped with a high-efficiency DC-DC converter to ensure the high efficiency and safety of the battery charging process.

In simple terms, the core of this technology lies in standardized storage modules. As a stationary energy storage unit in the rural PEDF system, these modules in the battery swap cabinet can absorb PV, balance the difference between peak and valley, and even provide power for electric agricultural machinery when required. For example, tractors and other large machinery can be powered in parallel by multiple storage modules, while low-power machinery (e.g., short-distance vehicles) uses fewer modules. Allowing for strong seasonal and intermittent use, when agricultural machinery and electric vehicles are not in use, these modules can be re-installed into the battery swap cabinet system as an indoor or village-level energy storage system, to consume rooftop PV and further improve energy efficiency. In this integrated battery swap cabinet system, the DC 48 V standard storage module is the key; it supports parallel expansion and hot plug, and is compatible with electric vehicles and agricultural machinery. In this way, low-speed electric vehicles and fully electric agricultural machinery will be more widely and efficiently applied in rural areas. Here is a further description of the interaction scenarios:

(1) Battery swapping: The farmer drives the electric agricultural machinery near the battery swap cabinet, puts the depleted battery pack from the agricultural machinery in the battery swap cabinet, and loads a fully charged new battery pack.

(2) Charge management: The depleted battery pack is removed and then charged in the battery swap cabinet. Based on the current energy situation and energy demand of the microgrid, the system arranges the charging timing and speed intelligently to maximize the use of renewable energy.

(3) Energy allocation: In the event of peak energy demand or PV power generation deficiency, the system can release energy from a fully charged battery pack to the microgrid to help balance supply and demand. This process is directed by the microgrid's central management system, thus ensuring the stability of the grid.

(4) Emergency response: In case of extreme weather or other emergencies, the battery swap cabinet can be quickly switched to emergency mode, and battery resources are allocated to provide a standby power supply for critical infrastructure.

In the new rural PEDF power system, the electric agricultural machinery is not merely the traditional agricultural tool, but a multi-functional energy unit to achieve efficient synergy of energy production, storage, and consumption. This model has significantly improved energy efficiency, and offered new perspectives and possibilities for sustainable energy development and smart agricultural practices in rural areas.

6.3.7 Summary of Technical Schemes in the Demonstration Village

To sum up, the topology of the new PEDF power system in Village A is as shown in Fig. 6.30, by considering the rural heating, cooking, RRB grid design, village-level microgrid, and energy storage solutions. The power system is divided into the indoor system and the transformer coverage area system, with a voltage level of 375 V. Allowing for the furthest distance (not more than 200 m) between RRBs in the transformer coverage area and the insignificant cable voltage drop effect, a higher voltage level is not selected for the busbar in the transformer coverage area. For the sake of economy, the indoor busbar and the busbar in the transformer coverage area are set to the same voltage level, which can reduce the busbar connection with DC/DC converters and reduce the cost. The indoor system includes PV panels, PV DC/DC converters, batteries, energy storage DC/DC converters, and DC/DC transformer converters for electrical appliances. The PV power is supplied to the indoor busbar through the PV DC/DC converter. The power of the busbar first meets the electricity demand of electrical appliances and indoor batteries, and the excess power will be supplied to the busbar in the transformer coverage area. The power of the busbar in the transformer coverage area gives priority to meeting the electricity demand for public equipment in the village and the charging demand for the storage equipment in the transformer coverage area, including the village collective canteen, ground source heat pumps, and vegetable greenhouses. The energy storage in the transformer coverage area is achieved by a battery swap cabinet with standardized batteries, and the batteries are used to meet the needs of electric agricultural machinery. The mode of battery swapping avoids the problem that electric agricultural machinery cannot provide energy storage for rural microgrids during busy farming. The whole power

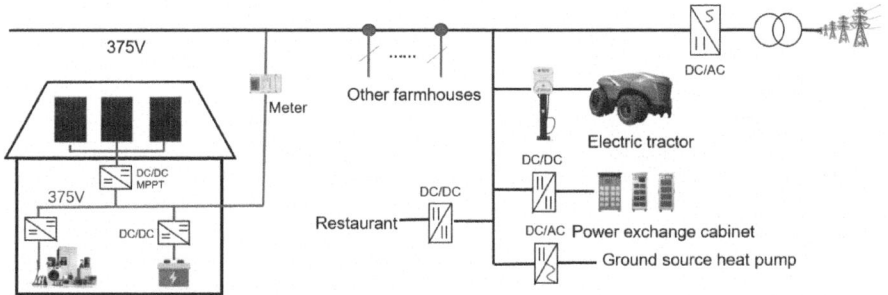

Fig. 6.30 Topology of the new PEDF power system in Village A

system is controlled by using the adaptive control strategy based on the busbar voltage signals. For details, see 6.2.

The new PEDF power system in Village A has the following characteristics: ① In terms of power system structure, the PV power is first supplied to peasant households for consumption, and PV power participates directly in rural energy use, thus guiding the full electrification of rural energy use. ② In terms of control methods, the distributed adaptive controller-free mode is employed to reduce communication and control costs. This is one of the important reasons for DC use. ③ In the mode of energy storage, the standard battery swap cabinet of electric agricultural machinery is used as the energy storage unit to promote agricultural electrification in the future, so as to provide "free" energy storage resources for rural areas. No centralized storage power sources will be built in rural areas. ④ In terms of performance parameters, the installed PV capacity is greater than the capacity of the transformer (at least 1.5 times), breaking the restrictions of transformers for rural PV systems.

6.4 Ways of Commercializing Rural Biomass Fuel

6.4.1 Consumption of Rural Biomass

(1) Current Status of Crop Straw Consumption in China

Based on the statistical data in Chap. 4, in 2021, the total quantity of crop straws in China was approximately 967 million tons, and with the utilization factor considered, this number was about 760 million tons. The main consumption methods for straw resources in China include the utilization of straws as manure, feed, fuel, base materials, and raw materials (referred to as the "five methods") as well as open burning, as shown in Fig. 6.31. Currently, nearly 200 million tons of straws in China are not effectively utilized or are discarded or burned in the wild.

Except for open burning, other consumption methods of straws have various degrees of positive contributions to its transformation into resources, however,

Fig. 6.31 "Five methods" of straw utilization in China and open burning of straws. **a** Utilization of straws as fuel; **b** Utilization of straws as manure; **c** Utilization of straws as feed; **d** Utilization of straws as raw materials; **e** Utilization of straws as base materials; **f** Open burning of straws

6.4 Ways of Commercializing Rural Biomass Fuel

they also have some problems. The advantages and disadvantages of different consumption methods of straws are summarized in Table 6.2.

Restricted by factors such as the degree of agricultural intensification, funds, and technology, China still has a relatively low comprehensive utilization of straw resources and still has the phenomenon of open burning of straws. Open burning of straws not only wastes resources but also damages farmland ecosystems and results in decreased soil fertility, deteriorated soil moisture content, and reduced soil microbial quantity and diversity. Furthermore, a large number of pollutants, such as $PM_{2.5}$, SO_2, NO_x, and CO, produced in the burning process (Table 6.3) can result in the deterioration of the atmospheric environment. According to a report by the United Nations Environment Programme (UNEP), open burning of straws can contribute greatly to the formation of smog, which affects local air quality and visibility and causes serious social impacts such as expressway closures and flight delays or even cancellations. Open burning of straws will also increase human exposure to pollutants, endanger people's health, and worsen diseases such as asthma. Therefore, to achieve reasonable consumption of straw resources, the first problem to be solved is the open burning of straws.

Open burning of straws has been strictly prohibited in various regions in China in recent years, which has achieved certain results under the efforts of all sides, however, open burning of straws persists in a considerable number of places in the absence

Table 6.2 Advantages and disadvantages of different consumption methods of straws

Straw consumption method	Advantage	Disadvantages
Utilization as fuel	Reducing fossil energy consumption and greenhouse gas emissions	Having a certain degree of combustion pollutant emissions and high storage and transportation costs
Utilization as manure	Improving the physical and chemical properties of the soil and increasing organic matter content, soil surface moisture, and the species and number of soil microbes	Increasing the risk of crop pests and diseases, possibly causing soil loosening, and resulting in loose root systems, weak seedlings, easy lodging in the later stage, etc.
Utilization as feed	Reducing breeding costs and improving milk and meat quality	High labor intensity, insufficient palatability and nutrients, and high storage and transportation costs
Utilization as base materials	High yield, with mushroom bran possibly returned to the field after production	Relatively high costs and certain technical thresholds
Utilization as raw materials	Being used to produce ethanol, building materials, etc., to alleviate problems such as food shortages and deforestation	High early-stage investment, high storage and transportation costs, and potential pollution such as wastewater
Open burning	Rapid treatment of straws	Wasting resources and causing serious pollution

Table 6.3 Emission factors of open burning of straws

Crop	SO_2 (g/kg)	NO_x (g/kg)	VOC (g/kg)	CO (g/kg)	$PM_{2.5}$ (g/kg)
Rice	0.1	2.3	4.0	52.1	3.0
Wheat	0.3	2.4	6.5	59.3	7.4
Corn	0.3	2.6	8.8	68.6	8.4

of supervision or even under strict supervision despite repeated bans. Therefore, to effectively curb the open burning of straws, the first thing to do is analyze the reasons why farmers choose to do so.

The most direct reason why farmers burn straws is that they want to quickly dispose of the straws left in the field after harvesting the crops so that they can soon sow seeds for the next season. However, the root cause for burning straws is that farmers choose this method with the lowest disposal cost to maximize their economic interests. In the past, farmers would collect straws for use as fuel or livestock feed when the rural economy was poor or there was a shortage of fuel. With the sustained and rapid development of China's economy, income levels have been greatly improved in rural areas, and commercial energy, such as coal and liquefied gas, has gradually been accessed by rural households. The continuous expansion of the urban–rural income gap, in particular, has caused most young and middle-aged rural laborers to choose to work in cities, making the opportunity cost of collecting and treating straws increasingly high. As a result, burning straws directly in the field has become farmers' "last resort". Therefore, the key to solving the burning of straws is to find a way for farmers to reasonably consume straws.

(2) Consumption Potential of Different Consumption Methods of Straws

Regarding the different consumption methods of straws, it is important to first consider whether they have sufficient consumption potential, i.e., the difference between their maximum consumption and their current consumption, to ensure the capacity of straws burned in the open air is completely consumed. The utilization of straws as feed, base materials, and raw materials is restricted by demand-side factors, such as the livestock population, the sales volume of edible fungi, and the usage of industrial raw materials, and the calculated maximum consumption of these methods is 240 million tons, 54 million tons and 59 million tons per year, respectively (Table 6.4). With these figures, the maximum additional consumption potential of the utilization of straws as feed, substrates, and raw materials can be obtained as 70 million tons, 14 million tons, and 39 million tons, respectively (whether these are the best consumption methods is not discussed here, which is discussed subsequently), which means a big gap with the current quantity consumed in open burning of straws of 180 million tons, indicating that the utilization of straws as feed, substrates and raw materials is insufficient to replace open burning of straws. Therefore, it is wishful thinking to expect that the problem of crop straw consumption in China can be completely solved through these three methods.

6.4 Ways of Commercializing Rural Biomass Fuel

Table 6.4 Maximum consumption and current consumption of different consumption methods of straws

Straw consumption method	Maximum consumption (100 million tons)	Current consumption (100 million tons)	Additional consumption potential (100 million tons)
Utilization as feed	2.40	1.7	0.70
Utilization as base materials	0.54	0.4	0.14
Utilization as raw materials	0.59	0.2	0.39

In contrast to the methods of utilization as feed, base materials, and raw materials with limited consumption, the utilization of straws as fuel can extensively replace the use of fossil energy which is consumed in a large amount every year in building heating, cooking, and industrial heat use in China, with almost no upper limit for its consumption potential, and it can completely consume straws burned in the open air and other straws consumed unreasonably. On the other hand, the utilization of straws as manure (returned to the field) can realize 100% consumption of straws by returning them to the field, and this method is chosen in many developed countries such as the U.S. to treat straws. Therefore, from the perspective of consumption potential, the latter two seem to have no differences. However, what are their impacts on the environment? A detailed comparative analysis is provided below.

(3) Return to the Field or Utilization as Fuel: Environmental Impact Analysis of Two Consumption Methods

Currently, the main method of utilization of straws as manure in China is the direct return of straws to the field, which is widely considered to have two greatest advantages: soil fertilizing and greenhouse gas emission reduction. From the perspective of soil fertilizing effects, the return of straws to the field can release nutrient elements such as N, P, and K and can theoretically play a role in fertilizing the soil and reducing chemical fertilizer use. However, in actual agricultural production in China, returning straws to the field has not yet played the role of replacing chemical fertilizers, for which there are two main reasons. The first one is technical reasons. Returning straws to the field to replace chemical fertilizers requires precision fertilization technology, i.e., adjusting the chemical fertilizer amount used in farmland based on the actual situation of the returned straws, however, precision fertilization technology is just getting started in China and its application cases are very few. The second is planting habit reasons. On the one hand, to ensure grain production, farmland in China has long been in a state of excess fertilization (as shown in Fig. 6.32, the chemical fertilizer application amount per unit area of farmland in China is far higher than the world average); on the other hand, the process of urbanization, as well as rapid social and economic development in China, have resulted in the reduction of young and middle-aged labors in rural areas and an increase in the opportunity cost of farming and that more and more farmers have begun to pursue more convenient planting ways, for

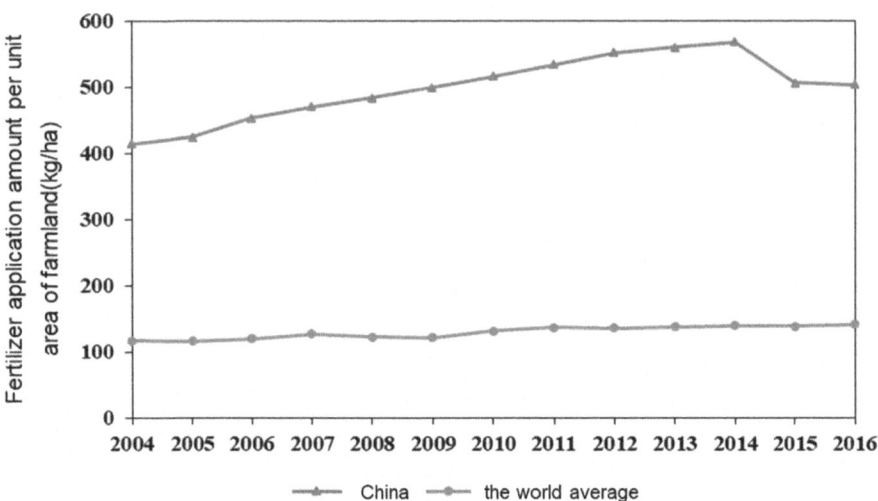

Fig. 6.32 Comparison of fertilizer application amount per unit area of farmland

example, in Northeast China where the authors conducted field experiments, when farmers sow corn seeds, they often conduct excess fertilization through the "one shot" method, to save additional topdressing during the elongation stage, thereby saving both time and labor. Therefore, the above reasons make it difficult to return straws to the field to replace chemical fertilizers at present. Furthermore, returning straws to the field is also faced with many other practical problems. For example, excessive straws returned to the field will absorb a large amount of original nitrogen, phosphorus, and water from the soil, which will lead to an unbalanced carbon–nitrogen ratio in the soil and affect the growth and reproduction of soil microbes, thereby affecting the returned straws' decomposition and conversion and crops' growth and development; when the straws are too long or the soil temperature is too low or compactness is inappropriate, it is to the disadvantage of straw decomposition and will affect seedling emergence and crop growth, leading to reduced production; returning straws to the field will also cause pathogens to return to the soil and long-term accumulation will aggravate seedling diseases and soil-borne diseases.

Most of the biomass contained in living plants will eventually be transformed into dead organic matter, some of which decomposes rapidly and reduces carbon to the atmosphere, however, some will remain for months, years, or even decades. Once the dead organic matter is broken and decomposed, it turns into soil organic matter which contains various substances that vary greatly depending on the duration they remain in the soil. Most of the substances consist of unstable compounds that are easily and quickly decomposed by microbial organisms to reduce carbon to the atmosphere, and only a small part of soil organic carbon turns into compounds that are difficult to decompose (such as organic–inorganic complexes) and are decomposed slowly, therefore remaining in the soil for decades to hundreds of years or even longer. According to some studies, the soil's carbon sequestration capacity is limited. When

6.4 Ways of Commercializing Rural Biomass Fuel

its carbon sequestration capacity becomes saturated, it will stop carbon sequestration, and from a very long time cycle, the soil's direct carbon sequestration ratio of biomass is almost negligible, at only about 3%. Plants absorb CO_2 during their growth and fix it in their bodies, and if all the carbon is released in the form of CO_2 again after the plants die and rot, it will be absorbed in a new round of plant growth, therefore, biomass can be considered to have "carbon neutrality" properties from the perspective of the conservation of matter. However, if certain quantities of gases with a strong greenhouse effect, such as CH_4 and N_2O, are released in the biomass consumption process, as the global warming potential of the two is 28 times and 265 times that of CO_2, respectively, the biomass will cease to be "carbon–neutral" and will increase the greenhouse effect. Different consumption methods are described below.

(1) The emission of CH_4 during the return of straws to the field is determined by a series of reactions of methanogens and methane-oxidizing bacteria. When the soil is in an anaerobic environment, such as the paddy field environment, the activity of methanogens is strong, which will greatly increase the emission of CH_4, and when the soil environment transforms from anaerobic to aerobic, such as the dry farmland environment, the activity of methane-oxidizing bacteria is strong, which will oxidize CH_4 and reduce its final emission; furthermore, nitrification and denitrification involving soil microbes will lead to N_2O emissions.

(2) Various kinds of biomass, whether agricultural waste or forestry waste, will lead to CH_4 and NO emissions during combustion due to chemical reactions, however, the emissions will vary significantly depending on different fuel forms and combustion methods.

(3) The process of biogas production through straw fermentation will produce a large amount of CH_4 gas. Ideally, the CH_4 gas will eventually be fully converted into CO_2 through combustion, however, if not well controlled, biogas may have leakage problems in the production and use processes similar to those in the extraction and use of fossil gas fuels (such as gas leakage in coal mining and leakage in natural gas extraction and transmission pipe networks), causing a part of CH_4 and N_2O gas molecules to be directly emitted into the atmosphere.

(4) In stockbreeding, CH_4 is mainly emitted in the gastrointestinal digestion process of various livestock species, among which the most CH_4 is produced and emitted during enteric fermentation in the digestive system of ruminants, and CH_4 and NO are also emitted during the fermentation of livestock and poultry manure.

(5) Even if mature straws are naturally discarded in the field without any treatment, they will eventually generate a certain amount of CH_4 and N_2O under the synthetic action of sunlight, rainwater, air, and microbes.

Based on research results in the literature, the CH_4 and N_2O emission factors for different biomass consumption methods are organized in Table 6.5.

Table 6.5 shows that all the consumption measures for straws and other biomass resources, including natural discarding, will emit a certain amount of CH_4 and N_2O in the entire process and result in varying degrees of equivalent incremental emissions, with returning rice straws to the field, utilization of straws as livestock feed, burning of

Table 6.5 CH_4 and N_2O emission factors for different biomass consumption methods

Consumption method	Emission factor [g/(kg of biomass)]		Equivalent incremental emission (gCO_{2e})
	CH_4	N_2O	
Returning corn stalks to the field	0.038	0.058	16.4
Returning rice straws to the field	40.5	0.002	1134.5
Combustion of straw briquette fuel	0.0067	0.0504	13.5
Burning of corn stalks in the wild	3.9	0.1	135.7
Burning of forestry waste in the wild	4.7	0.26	200.5
Biogas production through straw fermentation	3.866	0.001	108.5
Utilization of straws as livestock feed	17.92	0.213	558.2
Natural discarding of straws	0.0914	0.0802	23.8

forestry waste in the wild and biogas production through straw fermentation causing the largest equivalent incremental emissions. In contrast, the emissions of CH_4 and N_2O from clean and controlled combustion of straws processed and compressed into briquette fuel are the least among all the methods. It should be noted that currently, the greenhouse gas emissions of biomass during the entire life-cycle fermentation are generally underestimated. Related studies generally have a short period (usually one year or several years) due to the limitations of manpower, material, and financial resources while the decay and degradation of organic matter in biomass often take decades or even hundreds of years. However, even if the change process and the emissions of greenhouse gases such as CH_4 and N_2O of residual substances in the future long cycle are ignored, the phenomenon of incremental emissions has already been apparent. Therefore, it can be concluded that the shorter the internal carbon element of mature biomass resources is converted into CO_2 through clean and efficient combustion technology to serve as a carbon source for the next-round growth of crops, the easier to maintain the "carbon neutrality" of the biomass itself. Otherwise, whether the biomass resources are naturally discarded or otherwise utilized, the process will continue to emit CH_4 and N_2O and enhance the greenhouse effect.

The utilization of straws as fuel has become relatively mature after years of research and development. Both the collection, storage, and transportation costs and the combustion efficiency and emission performance of the processed straw fuel have been greatly optimized. Processing crop straws into fuel to replace fossil energy can realize real zero-carbon emissions, on the one hand, and achieve considerable pollutant reduction effects, on the other hand. Therefore, the utilization of straws as fuel is the best way to consume straw resources in China, however, it is important to achieve clean and efficient combustion to minimize the emissions of particles and other air pollutants.

6.4.2 Biomass Fuel Development Methods

Compared with wind energy, hydro energy, solar energy, etc., biomass energy is the most stable renewable energy, has the chemical energy storage property, and is more convenient for storage, transportation, and conversion. More importantly, with other renewable energy sources mainly used to output electric energy, biomass may be the only zero-carbon energy that can be used as fuel in the future, and biomass can be converted into diversified products such as combustible gas, heat, oil, steam, fertilizer, and materials, therefore, it should be valued. Currently, the main goal of utilizing biomass as fuel is to replace the role of traditional fossil energy while effectively consuming biomass straws.

The basic principles of biomass fuel development include one, transforming biomass energy development and utilization from passive consumption to active utilization; two, conducting classified and rational utilization of dry and wet biomass resources; three, building an economically sustainable "collection, storage, transportation and utilization" integrated industry chain, fully realizing the commercialization of biomass energy and increasing its added value, to attract social investment, increase rural income and contribute to rural revitalization.

Dry biomass raw materials mainly have three application scenarios: first, in localities very rich in biomass resources, electricity peak shaving for solar and wind power using biomass power plants (including combined heat and power) can be considered; second, biomass is transformed into commercial energy that is easy to measure and supply externally through compression molding or pyrolysis and gasification and is then burned in distributed dedicated boilers to replace small and medium-sized coal-fired boilers to provide heat sources for building heating or industrial steam heating; third, rural households, which have a bulk coal substitution need and cannot achieve clean heating through other green renewable energy resources, can use household clean biomass stoves to replace traditional rural bulk coal stoves, so as to achieve local internal consumption of biomass resources, and if there are still biomass resources remaining, they can be processed into commercial energy and then exported.

(1) Biomass Power Generation and Combined Heat and Power

From the perspective of biomass consumption only, biomass power plants undoubtedly can quickly dispose of massive crop straws, therefore, they have undergone fast development in some regions in recent years. As of the end of 2018, China had put 321 agricultural and forestry biomass power generation projects into production, with a total installed capacity of 8.063 million kW, an annual energy production of 39.47 billion kWh, and an average utilization hours of 4,895 h, of which the total installed capacity of combined heat and power was 3.46 million kW. From the perspective of energy grade, biomass power generation produces high-grade electricity, and combined heat and power units can also supply heat. However, despite high energy efficiency, most coal-fired units in China have been conducted or are undergoing ultra-low-emission transformation, and if biomass power plants are further used to replace coal-fired power plants, it will require repeat investments and can only bring small environmental benefits.

Currently, the mainstream installed capacity of units in biomass power plants is 30,000 kW, which requires large fuel consumption and storage. However, the dispersed distribution of biomass resources will expand the radius of raw material collection and make transportation difficult. In particular, in autumn, the harvest season, large amounts of biomass fuel are intensively transported to the plants, which results in road congestion and increased transportation costs, and with high raw material costs, biomass power plants become overly dependent on the government's subsidies for electricity prices, which is a core issue restricting their development. Biomass power plants build biomass raw material inventories based on 15 to 70 days of use quantity. Taking a biomass power plant with an installed capacity of 30,000 kW as an example, it needs to store nearly 50,000 tons of raw materials, which makes fire management difficult and occupies a large amount of the enterprise's operating cash flow. Furthermore, due to the mismatch between the heat source and the heating load, a large part of the heating capacity of biomass combined heat and power projects is left unused. It is difficult for newly built biomass combined heat and power projects to find heating load demands in cities and towns that match their heating capacity, which causes the units to operate in the pure power generation mode, and no mature distributed small-scale combined heat and power systems are available to supply heat to villages and towns. As a result, the situation of mismatch between the heat source and the heating load cannot be broken.

From a macro perspective, as of the end of 2018, the installed capacity of renewable energy reached 755.81 million kW in China, of which the installed capacity of wind and solar power reached 358.89 million kW, with serious wind and solar curtailments. Therefore, if the construction of biomass power plants is blindly promoted, the energy eventually replaced will not be fossil energy but the energy production by existing renewable energy sources, such as wind and solar power, which will waste the clean attribute of biomass resources or even contribute negatively to air pollutant emission.

(2) Small and Medium-sized Biomass Boilers

Regional central heating with biomass mainly involves the use of small and medium-sized biomass briquette fuel boilers or boilers for directly burning bundled straws, the design capacity of which should not be too large subject to the constraints of collection radius, storage scale, and fuel cost. Such boilers can be mainly developed in regions with rich biomass resources, severe air pollution, or a heavy task of eliminating coal-fired boilers, public facilities, and commercial facilities with much bulk coal consumption, such as county schools, hospitals, hotels, and office buildings, as well as populated areas such as rural towns, or they can be applied in small and medium-sized industrial parks as well as industrial areas not covered by central heating, to provide clean and economical industrial steam to industrial users and reduce fossil energy consumption. Small and medium-sized biomass boilers used for central heating can bring into full play the zero-carbon property of biomass and replace a large amount of coal consumption, and the flue gas of the boilers can be centrally treated to meet emission requirements.

(3) Heating Through Pyrolysis and Gasification of Biomass

6.4 Ways of Commercializing Rural Biomass Fuel

Biomass gasification is the process of converting hydrocarbons that makeup biomass into combustible gases, such as carbon monoxide, hydrogen, and low-molecular hydrocarbon gases (such as methane), under certain thermal conditions.

Based on gasification equipment and running status type, it can be divided into fixed bed gasification and fluidized bed gasification. The gas composition and calorific value of different raw materials are shown in Table 6.6.

(4) Household Clean Heating and Cooking with Biomass

Biomass resources have many sources, and the most commonly used are agricultural and forestry biomass materials, mainly including bark, leaves, wood, and straws. If not processed, these raw materials are not convenient for large-scale utilization and their combustion efficiency is low and pollutant emissions are high due to their loose structures, dispersed distribution, transportation and storage inconvenience, low energy density, and irregular shapes. Biomass briquette fuel processing technology squeezes agricultural and forestry residues, such as crop straws, rice husks, branches, bark, and wood chips, into pellets or briquettes with specific shapes and high density through special kneading and cutting (crushing), drying, and compression equipment. Such technology is key to the efficient utilization of biomass. Compression molding technology can largely increase the density of biomass so that the briquette fuel has an energy density equivalent to that of coal with medium calorific value, has transportation and storage properties similar to those of coal, has high efficiency and low release of pollutants when it is burned in cooking or heating furnaces, and can replace conventional fossil energy such as coal, liquefied gas and natural gas, to meet household energy demand in life, such as cooking, heating and domestic hot water. Encouraged and supported by the national clean heating policy in recent years, some places in China have made attempts to replace bulk coals with new biomass briquette fuel stoves and conduct clean heating.

(E) Biogas-Based Natural Gas Engineering

Biogas-based natural gas technology produces biogas-based natural gas by utilizing agricultural and other organic wastes, and turns low-grade biogas into high-value-added high-grade energy—biogas-based natural gas—through purification technology. The technology expands the biogas scale, enables it to have economies of scale, and increases its efficiency while achieving professional management so that it has broad prospects for utilization. The technology also returns biogas residue and slurry to crops for use, to make the entire system form a closed loop and realize sustainable continuous full-industrial chain development.

Table 6.6 Gas composition and calorific value of different raw materials

Raw material	CH_4(%)	CO_2(%)	CO(%)	H_2(%)	O_2(%)	Calorific value (kJ/m^3)
Rice husk	2.60	13.50	14.50	9.50	1.61	3762
Wood chip	3.94	19.62	12.62	12.35	0.27	5300
Apricot shell	3.32	17.64	10.84	14.87	0.27	4351

The components of biogas produced by anaerobic digestion include CH_4 (50–65%), CO_2 (30–38%), N_2 (0–5%), H_2 (<1%), O_2 (<0.4%), and H_2S (500 ppm) and a certain amount of moisture. Purified biogas should meet the national standard for natural gas as vehicle fuel, *Compressed Natural Gas as Vehicle Fuel* (GB 18,047–2017), including higher heating value >31.4 MJ/m^3, $H_2S \leq 15$ mg/m^3, $CO_2 \leq 3.0\%$ and $O_2 \leq 0.5\%$. Biogas produced by anaerobic fermentation is purified to extract CH_4 which is injected into the natural gas pipe networks or used for vehicles.

The *Guiding Opinions on Promoting the Industrialization of Biogas-based Natural Gas* jointly issued by 10 authorities in December 2019 proposes that China's annual output of biogas-based natural gas should exceed 10 billion m^3 by 2025 and exceed 20 billion m^3 by 2030. Since 2015, the National Development and Reform Commission and the Ministry of Agriculture and Rural Affairs have invested RMB 2 billion annually to support 64 large-scale biogas-based natural gas pilot projects and more than 1,400 large-scale biogas projects for three consecutive years, and as of the end of 2020, China had completed a total of 55 biogas-based natural gas projects, with an annual output of 211 million m^3 of biogas-based natural gas.

6.4.3 Biomass Energy "Production-Supply-Sales-Service" Trading System

Currently, as the only zero-carbon fuel, biomass can replace various fossil fuels and is expected to become a main commodity in the future fuel market. All commodities should be circulated and should participate in market competition. The traditional fossil energy trading model should not be directly adopted for biomass energy that represents an emerging industry, but a "production-supply-sales-service" integrated trading system in line with its characteristics should be built.

The "production-supply-sales-service" integrated trading system can efficiently integrate the production, transportation, storage, sales, and service of biomass fuel products, minimize intermediate links, interact with, mutually help, and achieve coordinated development with upstream producers, midstream dealers, and downstream consumers, to improve circulation efficiency, reduce operating cost and achieve the goal of profiting together.

(1) Mode of Distributed Processing and Production of Biomass Pellet Fuel

The new "one village, one factory" mode of processing and production of biomass pellet fuel is adopted to avoid the drawbacks of the existing operating mode of large-scale concentrated processing of biomass in China, such as uncontrollable raw material prices, high transportation costs due to large collection radius and high finished product prices due to multiple circulation links. The mode operates in the following way: the villagers' committee is responsible for and manages the biomass pellet production factory, hires local people to conduct production, operation, and maintenance, and requests rural households to send their fruit tree branches to the processing factory for processing during autumn and winter when they prune fruit trees and requests them to transport the finished product back to their respective homes after the branches are processed into fuel. In this process, rural households only need to pay a small amount of processing fee to subsidize necessary expenses such as the base pay (accounting for about 40%) of production workers, equipment electricity fee (accounting for about 50%), and maintenance cost (accounting for about 10%), and the factory runs six to eight hours a day and conducts concentrated processing for three to four months, which can produce 200 tons of fuel in total for local cooking for the whole year. Rural households that have many raw materials can consider selling them to the villagers' committee at a certain price, and then the villagers' committee can uniformly sell the fuel to users with larger demand in the village or users in neighboring villages, to not only improve equipment utilization efficiency and help maintain the normal operation of the factory building but also further promote the development of local and neighboring villages in renewable energy utilization and expand the publicity effect.

(2) Cloud Platform-Based Biomass Fuel Terminal Service Mode

The cloud platform is a real-time management system that integrates the advantages of informatization, networking, intellectualization, and integration and is also a new operating mode. The core elements include the open interface, real-time feedback output, and dynamic matching, and the information density and data scale are not only important goals for cloud platform building but also important bases for performance evaluation.

The cloud platform's module with high practical value is the real-time and dynamic supply–demand information matching mechanism for the production, supply, sales, and service of biomass fuel. Via data access ports, diversified information, such as rural households in the places of origin, enterprises, market demands, prices, and end users in the front end of the biomass industry chain, gathers on the platform. The sending, transmission, and feedback of information are completed in real-time and presented to customers based on permissions.

Open Access This chapter is licensed under the terms of the Creative Commons Attribution-NonCommercial-NoDerivatives 4.0 International License (http://creativecommons.org/licenses/by-nc-nd/4.0/), which permits any noncommercial use, sharing, distribution and reproduction in any medium or format, as long as you give appropriate credit to the original author(s) and the source, provide a link to the Creative Commons license and indicate if you modified the licensed material. You do not have permission under this license to share adapted material derived from this chapter or parts of it.

The images or other third party material in this chapter are included in the chapter's Creative Commons license, unless indicated otherwise in a credit line to the material. If material is not included in the chapter's Creative Commons license and your intended use is not permitted by statutory regulation or exceeds the permitted use, you will need to obtain permission directly from the copyright holder.

Chapter 7
Key Technologies for Low-Carbon Development of Rural Energy

Rural areas in China possess zero-carbon resource conditions, energy use characteristics, and development prospects for achieving the energy revolution. This requires technical support in terms of energy development, energy application, promotion mode, etc. This chapter introduces how to achieve the above goal and key technologies.

7.1 Rural House Rooftop PV Installation Technology

Rural house rooftop PV system installation is a technically complex task, with the main installed components as shown in Fig. 7.1, and multiple factors need to be considered. The installation process of a rural house rooftop PV system is presented in this section from aspects such as the pre-evaluation of roof bearing capacity, etc., precautions during installation, and maintenance during operation.

To add or transform a PV system on a rural house, an on-site survey should first be conducted to evaluate the roof bearing capacity and area as well as the building structure safety and PV system electrical facilities which should meet roof waterproofing, windproofing, lightning protection, fire protection and anti-static functions and building energy efficiency requirements. When evaluating the roof-bearing capacity, the following factors need to be considered. ① Roof material: Different types of roof materials, such as concrete, tiles, and metal, have different strength and durability characteristics, and it is important to fully ensure their support of the weight of the PV system. ② Weight capacity: The weight capacity of the roof is another key factor. Furthermore, the locations of beams in the roof structure should also be considered to ensure that the roof can bear the weight of the PV system. Reinforcement measures should be taken if necessary.

Based on the pre-evaluation of the roof bearing capacity, the installation method and paving area of the PV panels should be selected according to the actual situation

Fig. 7.1 Rooftop PV installation components

of the roof in the installation process, which can be considered from the following factors. ① Roof area: Determine the quantity and arrangement method of the PV panels by measuring the available area of the RRB roof. ② Tilt angle selection: The tilt angle of the PV panel will affect its energy harvesting efficiency, which in turn affects the power generation capacity of the system. The tilt angle of PV panels is generally between 15° and 40° in rural areas. The selection of tilt angle should be comprehensively considered according to factors such as local latitude and longitude, sunshine, and seasonal changes, to achieve the best power generation effect. ③ Installation direction: The installation position of the PV module should not lower the sunshine standards of the adjacent building or the building where it is located, and it should be preferentially installed on the south-facing roof or flat roof to maximize the energy production of the PV system. Furthermore, during the actual installation, covers, and constraints from the surroundings should also be considered, such as trees, buildings, hills, or other obstacles, and the appropriate PV panel area should be selected to reduce the impact of shadows on the system. ④ Distance between modules: The spacing between modules should be reasonably arranged when installing PV panels. Appropriate spacing can ensure the ventilation and heat dissipation between the PV panels and avoid mutual shading that affects the power generation effect. Appropriate spacing is also convenient for future maintenance and cleaning of the PV panels.

The installation process is somewhat different depending on the roof material and structure in rural areas. For concrete roofs, the installation order is as follows: precast concrete counterweight construction, PV support installation, PV module installation, electrical equipment installation, cable laying, lightning protection and grounding, system commissioning, and system power generation. The support for a cement flat roof can be divided into two parts: the base of the support and the support itself. The base is the precast concrete counterweight of which the size, concrete strength, and arrangement position shall meet the drawing design requirements. The

7.1 Rural House Rooftop PV Installation Technology

precast concrete counterweights shall meet the design drawing specifications, and concrete test cubes shall be reasonably left based on the number of counterweights on the roof. The construction personnel shall not cause damage to the original structure of the roof. The support itself is generally an aluminum alloy support. For prepainted steel plate roofs, the installation order is as follows: waterproof treatment, fixture and keel installation, PV module installation, electrical equipment installation, cable laying, lightning protection and grounding, system commissioning, and system power generation. The common types of prepainted steel plates are mainly divided into the standing seam type, corner fitting type, bayonet type, and exposed nail type. Different types of prepainted steel plates need to be equipped with different mounting fixtures.

Rooftop PV systems should be regularly maintained to ensure their normal operation and extend their service life. The later-stage maintenance of rooftop PV systems in rural areas can be considered from the following aspects. ① Ensuring personnel safety: A rural house installed with a rooftop PV system should have safety protection measures in place for installation, operation, and maintenance as well as safety protective facilities that prevent components from falling after the PV panels are damaged. ② Waterproof measures: During rooftop PV system installation, waterproof measures should be taken to prevent rainwater from seeping into the interior of the building. PV panel installation should be ensured to cause no damage to the roof waterproof layer and appropriate waterproof measures should be taken, such as using waterproof adhesive tape or sealant to seal the interface between the PV panels and the roof. ③ Cleaning and maintenance: PV panel cleaning and maintenance is crucial for ensuring the long-term performance of the system. Dust, dirt, and other impurities on the surface of the PV panels will reduce the system's power generation efficiency, therefore, the panels should be regularly cleaned to maintain the system's best performance. The cleaning frequency can be determined based on local environmental conditions and seasonal changes. It is also important to regularly inspect various system modules, including the panel, cable, junction box, and inverter, to ensure their normal operation. Maintenance should be conducted and damaged or aged components should be replaced if necessary. ④ Performance monitoring: Monitoring the PV system performance can ensure timely identification and solution of potential problems and efficient system operation. For example, monitoring the system's energy production enables the evaluation of system performance and efficiency and the detection of any anomalies; as high temperature will affect the PV panel performance, monitoring the panel temperature can help identify potential overheating problems. The following should be calculated and monitored for rooftop PV systems in rural areas: the power generation system's installed capacity, annual energy production, outdoor temperature, solar energy global irradiation, rural house power consumption, and battery storage capacity.

Besides considering factors such as roof bearing capacity, PV installation, and operation and maintenance, rooftop PV systems in rural areas should also meet the following requirements. ① Rooftop PV systems in rural areas should be comprehensively utilized throughout the year, to supply electricity, domestic hot water, heat, or cooling to buildings based on the land climate characteristics, actual needs, and applicable conditions. ② Rooftop PV systems in rural areas should be prioritized for rural households, with the surplus electricity connected to the grid. ③ Rooftop PV systems in rural areas should be preferentially advanced in whole villages to form a rural microgrid system, to reduce the impact of unstable single households' connection to the grid on the grid.

7.2 Rural Microgrid Control Technology

Rural buildings have the characteristics of "large quantity and dispersed distribution", with the distance between different RRBs possibly reaching hundreds of meters. If the traditional centralized control method is adopted, it will greatly increase the controller and communication system costs and the maintenance cost of system faults will be high. Furthermore, the living habits of villagers vary tremendously. Some RRBs are uninhabited all year round, some RRBs only have power demand on weekends and holidays, and some RRBs may have relatively stable loads, which also increases the difficulty of centralized control. Therefore, the rural power system needs a decentralized adaptive control scheme, which requires the control system to have a simple structure, controllable cost, and clear control logic. Voltage is the simplest and most important operating parameter for DC systems, and using voltage signals as control signals to change the equipment running state is the basic idea of DC voltage signal control, as shown in Fig. 7.2. DC systems allow the voltage to fluctuate in a large range to ensure the system runs within a safe voltage range and meets system emergency response requirements. As the ultimate system running goal, steady-state control makes the system run in the target working area through the coordination of DC converters.

Specifically, power electronic equipment in the system (such as inverters, DC/DC converters, and power connection systems (PCS)) senses the voltage signals at the points of inductive contact and changes its running status based on the voltage level.

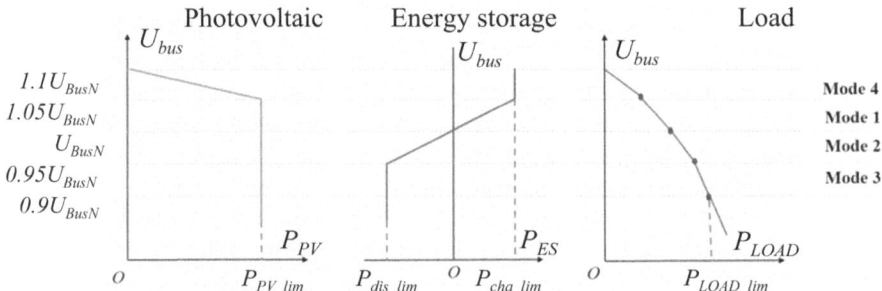

Fig. 7.2 DC busbar voltage control principle

7.2 Rural Microgrid Control Technology

This is a decentralized controller-less control method, in which the equipment makes adaptive adjustments based on the voltage signals of its location to maintain stable system running. Taking the PV DC converter as an example, when the voltage in contact is higher than the set voltage, the converter operates in the maximum power point tracking (MPPT) mode; when the voltage in contact is lower than the set voltage, the converter operates in the constant voltage mode. Similarly, the battery's DC converter also senses voltage signals at the point of contact, thereby controlling the possibly consumed busbar electric power; when the busbar voltage is relatively low, each set of equipment will change its running status, to ultimately increase the power injected into the bus and minimize the consumption of the busbar's electric power. In the specific operation, the voltage control strategy distinguishes the system running status by voltage range. After the system is overall started, it is divided into the operating battery's charging/discharging state and power. The control baseline of the entire system is the busbar voltage. To ensure the stable status and working status of the system: the running status is the target working area set by the system, allowing the voltage to operate stably on upper limit 1 of the running status, and the busbar voltage needs to be maintained relatively stable (allowed to fluctuate in a controllable range); when the busbar voltage needs to be relatively higher, each set of equipment will change its running status, to ultimately reduce the power injected into the busbar, make the voltage reasonably fluctuate between the upper and lower limits and weaken the information processing level; the working status is the lowest required status for keep equipment running, and equipment can still maintain certain running between the upper and lower limits of the operating voltage; exceeding the upper limit of the operating voltage will threaten equipment safety, and if it is lower than the lower limit of the operating voltage, the equipment will stop running.

In the grid-connected running status, the bidirectional AC/DC converter acts as a voltage regulating unit responsible for maintaining system power balance and DC voltage stability. The MPPT mode is used to control the PV interface converter to maximize solar energy utilization. When the system's internal energy supply exceeds the demand, the energy storage unit will preferentially absorb the excess energy, and when the capacity of the energy storage unit reaches the threshold, it will stop working, and the excess energy will be sent to the grid; when the energy supply falls short of the demand, the energy storage will preferentially release energy and meet the energy demand of the load together with the PV converter, and if the load demand still cannot be met, the insufficient energy will be supplied by the grid. When energy is transferred from the AC side to the DC side, the bidirectional AC/DC converter operates in the rectification status; on the contrary, when energy is fed back from the DC side to the AC side, the bidirectional AC/DC converter operates in the inversion status.

When the large power grid has a fault or is under scheduled maintenance, the DC distribution system changes to grid-independent operation. The bidirectional AC/DC converter withdraws from the operation of the DC distribution network, which will cause the voltage of the DC busbar to fluctuate. When the DC voltage deviates to a certain extent, the energy storage interface converter changes to the constant DC voltage mode, and the energy storage unit takes over the DC voltage to maintain

Table 7.1 Coordinated control strategy for converters under different operating modes

System running status	PV interface converter	Energy storage interface converter	Bidirectional AC/DC converter
Grid-connected operation	MPPT MPPT control	Power control	Constant DC voltage control
Grid-independent operation	/MPPT control/limited power	Constant DC voltage control	
Night operation		Power control	Constant DC voltage control

a new stable point. When the system's internal energy supply exceeds the demand, the PV interface converter operates in the MPPT mode, and the energy storage unit absorbs the excess energy. If the energy storage capacity reaches the upper limit, the PV interface converter will change to the power limit mode to reduce energy output by the PV array. When the system's internal energy supply falls short of the demand, the PV interface converter operates in the maximum power tracking mode and the load is jointly powered by the energy storage unit and the PV system. If the load demand still cannot be met, the power supply will be supplemented by the rectifier, and the unnecessary load can be reduced appropriately.

When operating in the night mode, the PV array does not output power, the bidirectional AC/DC converter operates in the constant DC busbar control mode (rectification mode) to supply power to the load and charge the energy storage unit.

The coordinated control strategy for converters under different operating modes is shown in Table 7.1.

When the voltage reaches the upper limit of the system operating voltage, the PV system conducts solar curtailment to keep the voltage from exceeding the upper limit of the operating voltage; when the PV system's power supply is insufficient and the voltage is lower than the lower limit of the system operating voltage, rectification is started to maintain the system in the normal operation interval. Between solar curtailment and rectification, different voltage gradient values are set based on the system operating mode and specific situation. Through the above method, the new power system in rural areas can realize decentralized group intelligent control, without the need for special controllers or communication equipment, and each converter automatically regulates its running status by sensing the busbar voltage of its location to achieve the entire system's stable operation.

7.3 PV Heating Technology for Rural Areas

Heating energy consumption is an important part of energy consumption in rural areas in northern China and a key issue to be solved by the new power system in rural areas. As RRBs are dispersed and rooms have different usage patterns, different PV heating schemes should be adopted for different types of rooms based on their usage patterns.

7.3 PV Heating Technology for Rural Areas

The difficulty for PV heating technology to overcome is the mismatch between PV power and room load, as shown in Fig. 7.3. Specifically, PV power is fluctuating while the heating load of the room is relatively stable. How PV power can meet the heating demand of the room at any time, especially at night and on overcast and rainy days, is a huge challenge. Another difficulty for PV heating is how to avoid the overheating problem in the afternoon on sunny days, especially in the initial and last cold periods, on the premise of meeting the heating demand of rural households. It should also be noted that rural households have limited economic levels and equipment costs should be within their acceptable range. Therefore, the PV heating technology for RRBs should have the characteristics of low cost, adjustability, and a high guaranteed rate. In the text below, rooms in RRBs are divided into those used all day and those used intermittently, and suitable PV heating technology is presented, respectively.

Taking the bedroom heating of an RRB as an example, a bedroom is used all day long, with a relatively stable heating load, in which heating is needed all day. However, the PV power is zero at night and on overcast and rainy days, and is difficult to meet the heating demand of the room. Therefore, the heat generated during the daytime on sunny days needs to be transferred for use at night and on overcast and rainy days to match the heating load characteristics of the room. Such a heat transfer process can be realized through energy storage. However, the cost of power storage is too high for rural households to afford. Heat storage technologies involving buried pipes and water tanks also have a high cost, and they occupy a large area, therefore, such technologies are difficult to implement in RRBs. As a result, walls are used for heat storage from an economical perspective. Red bricks are cheap and cover a small area, basically not compressing the living space of rural households. PV direct-driven heating wires are used for the heating terminal, and the control of voltage or power stability is not needed during the operation of the heating wires, which reduces the cost of equipment control.

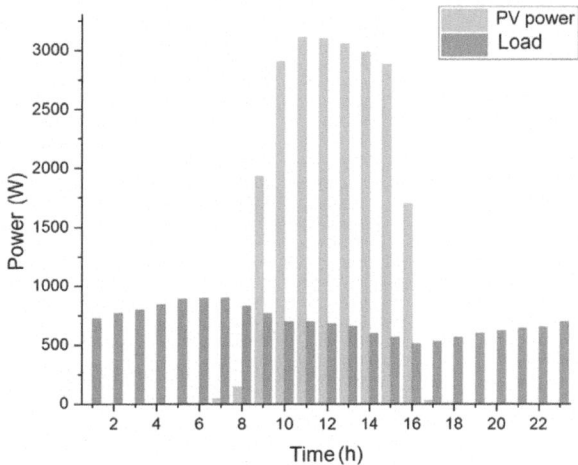

Fig. 7.3 Mismatch between PV power and rural house heating load

The structure of the heating system is shown in Fig. 7.4. The heating system has three parts: ① Power generation system that mainly consists of PV panels and DC/DC converters; ② Heating terminal that mainly consists of heating wires and insulation boards (such boards are not separately installed for buildings with insulation treatment); and ③ Heat storage system that can be a heat storage wall or other heat storage materials. The system operation process is as follows: Under the control of the converters, the PV panels generate power at the maximum power point, the current flows to the heating wires via the converters, the heating wires generate heat and then heat the inner surface of the heat storage body through heat conduction, the temperature of the heat storage body continues to rise under the action of heat conduction, and then heat is sent to the room from the outer surface of the heat storage body through convective heat transfer and thermal radiation. The insulation boards have the functions of reducing heat loss and making the heat of the heating wires flow to the heat storage body instead of the exterior walls. The PV panels generate power at the maximum power point throughout the process, with all the power generated used for the heating wires' heat generation, therefore, there is no solar curtailment, and the PV self-consumption rate is 100%. However, the PV installed capacity should be obtained based on the room load and through rigorous calculation, a constraint condition of which is that the PV energy production during the day should meet the heating demand of the room all day long and the maximum temperature of the heating wires should be lower than the safety temperature of the insulation boards.

An application demonstration was conducted for the heating system in a village in Shanxi Province in 2022, with the results of a one-week test in the RRB as shown in Fig. 7.5. It can be seen from the figure that the heat storage wall effectively eased the volatility of PV power and transferred heat from the daytime for use in the nighttime, thereby ensuring the heating demand at night. The heat storage wall made

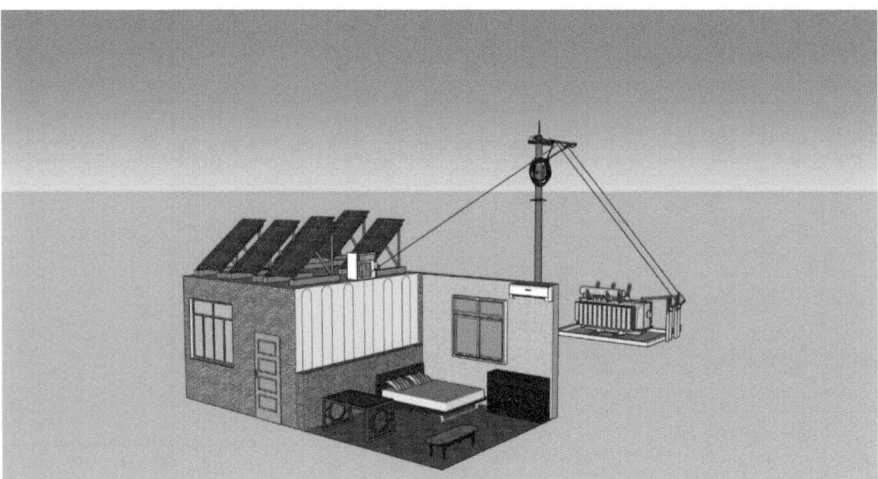

Fig. 7.4 Schematic diagram of PV heating system based on wall heat storage

7.3 PV Heating Technology for Rural Areas

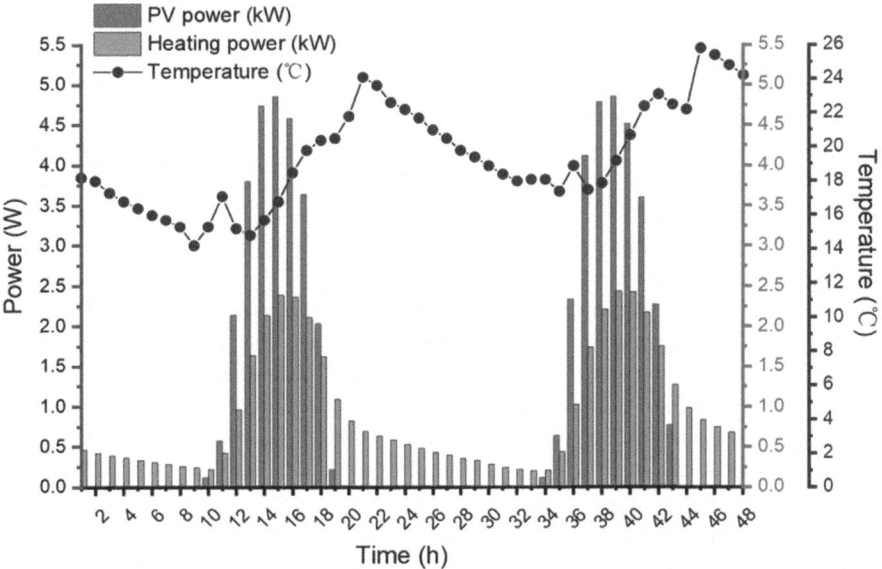

Fig. 7.5 Heating test effect*

the indoor temperature reach its highest around 7:00 PM and avoided overheating the room when the sunshine was most abundant in the afternoon. The weather conditions during this week included overcast, cloudy, and sunny conditions, and under different weather conditions, the average indoor temperature of the RRB was around 17 °C, and the indoor temperature could be maintained above 15 °C even at night to meet the heating demand of rural households. Furthermore, in the entire heating season, for more than 70% of the time, the indoor black globe temperature of the RRB room remained above 16 °C. The heat released by the heat storage wall at night accounted for 40% of the PV energy production. Therefore, the heat storage wall could effectively mitigate the supply–demand mismatch in the PV heating system and meet the heating demand of indoor personnel at any time.

Through the heating method based on wall heat storage, the PV heating system achieves the goal of low-cost heat storage, however, this method is more suitable for new buildings as the labor cost of transforming existing buildings is high and the construction is tedious. For existing buildings, heat can be supplied by using PV-driven heat storage electric heaters. Such heaters include, from the inside to the outside: the heating ceramic tube, refractory brick, insulation cotton, and equipment enclosure. The ceramic tube has two power connectors, one of which is connected to the PV power to drive the ceramic tube to generate heat, with the PV system always running at the maximum power point under the regulation of the converter, and the other of which is connected to municipal power to drive the electric heater in case of continuous extreme weather, to ensure indoor heating. The heat emitted by the ceramic tube first heats the refractory brick which can withstand higher temperatures and thus has stronger heat storage per unit mass than the red brick used in buildings. The heated refractory brick then heats the air in the sandwich of the electric heater through convective heat transfer, thereby supplying heat to the room. When there is no PV power at night, the heat stored in the refractory brick can still meet the heating demand of the room. The insulation cotton in the electric heater has the function of preventing the heat of the refractory brick from being conducted out from the surface of the electric heater. If it is conducted out, it will cause the room to overheat, and an excessive surface temperature of the electric heater will result in a safety hazard to personnel in the room.

Residents with high requirements for heating flexibility and temperature control can choose the PV-driven air-to-air heat pump to heat the room. The biggest technical challenge to the PV-air-to-air heat pump system is the maintenance of stable voltage on the power supply side so that the power of the system can meet the demand of the air-to-air heat pump at any time. In the meantime, the capacity configuration of the PV-air-to-air heat pump is an optimization problem, with the PV capacity and storage capacity greatly affecting the economy of the system. Therefore, control and optimization constitute the core technologies of the PV-air-to-air heat pump system. A Tsinghua University team conducted a PV-air-to-air heat pump demonstration in a village in Fangshan District, Beijing in November 2022, as shown in Fig. 7.6. The scheme adopted an adaptive control strategy based on busbar voltage signals, to stabilize the system voltage at 220–240 V. According to the test results, the PV power self-sufficiency of the air-to-air heat pump reached more than 70% during the heating period.

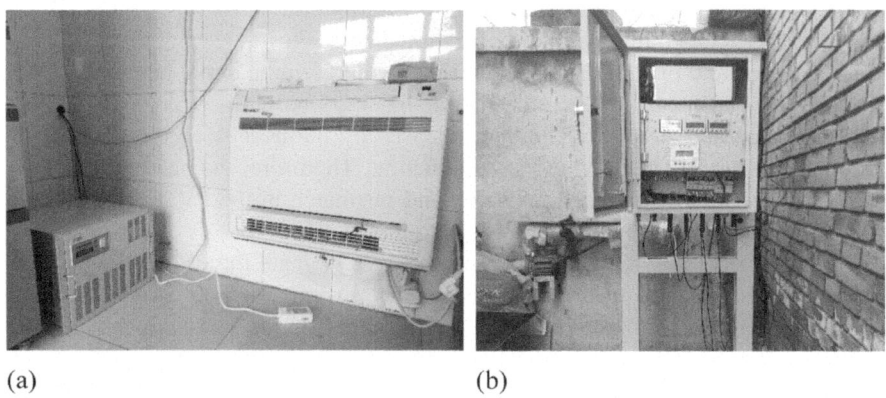

Fig. 7.6 PV-driven air-to-air heat pump system **a** Air-to-air heat pump indoor unit; **b** Control system cabinet

7.4 Isolated Grid PEDF System Technology

According to the latest data statistics from the World Bank, 840 million people in the world do not have access to electricity, a basic necessity of life, and are mainly distributed in Sub-Saharan Africa, parts of South Asia, parts of South America, and small parts of East Asia. China has 450 inhabited islands and more than 106 border towns. Rural residents in these remote areas face the problem of power shortfall, most of whom live a primitive farming life.

One of the important means to solve the power supply problem for people without access to electricity in the world is the construction of isolated grid systems which have advantages such as flexibility, energy conservation, reliability, environmental protection, low investment, and fast coverage compared to the traditional power grid. A PEDF isolated grid system, which takes the isolated grid system as its core and is integrated with the PEDF technology, has been proposed. This system consists of distributed generation units, energy storage equipment, and DC loads, solves power supply and utilization in rural remote areas, and strongly supports the rapid development of distributed energy of different scales, marking a major transformative technology for distributed energy utilization.

The countryside has characteristics such as rich roof resources and high compatibility with PV applications. Developing self-sufficient isolated grid PEDF systems in the countryside is highly economical and applicable and is a feasible way to achieve the distributed energy revolution, build a new power system, and realize the dual carbon goals in the countryside. The isolated grid PEDF mode can guarantee the supply of clean, safe, reliable, intelligent, and efficient electric energy, lead the energy power transformation of isolated grids, and enable everyone to access clean power.

7.4.1 Introduction of Isolated Grid PEDF System

It is very expensive and difficult for high-altitude areas, sea islands, border areas, and uninhabited areas, which are generally far away from centrally inhabited areas, to connect to the large power grid. The isolated grid PEDF system is equipped with solar photovoltaic panels to realize energy supply as well as electric energy storage of the energy storage system across hours. The system realizes source-load connection through DC supply and utilization and uses standby power supplies such as hydrogen fuel cells to achieve self-sufficiency in electric energy. Its layered busbar architecture adopts voltage levels of DC 800 V, DC 400 V, and DC 48 V, and depending on the voltage demand of the load, high voltage realizes efficient power consumption, and 48 V low voltage realizes intrinsic safety in power consumption.

The isolated grid PEDF system is ideally suited for farmhouses, rural individual livestock farms, mountain livestock farms, mountain temples, environmentally friendly agricultural projects, and communication base stations far away from living quarters, exploration team stations, and mountaintop outposts, where the system is installed very close to the power consumption locations but far away from the power grid.

7.4.2 Isolated Grid PEDF System Application

(1) PV Station in Holy Elephant Gate, Nam Co.

The project is in the Holy Elephant Gate scenic area, Chadolangka Island, North of Holy Lake Nam Co, Qinglong Township, Baingoin County, Nagqu, Xizang, with the lake at an altitude of 4,700 m. Mainly using containers as the carrier for space design (Fig. 7.7), the PV station system in Holy Elephant Gate, Nam Co is a source-storage-grid-load integration, creates a clean, safe, reliable, intelligent, and efficient living environment and adopts a grid-based layered architecture, constant/limited power flexibility control technology as well as constant/limited power flexibility operation technology with ET/QT situation awareness based on energy information. It can achieve rapid deployment and grid-independent operation. The station's PEDF system is not connected to the power grid, and it operates solely on an isolated grid. The rooftop is equipped with 2.46 kWp PV modules and an energy storage system of 3 kW/6.6 kWh. The system has multiple DC output circuits to meet the power demand of DC air conditioning, DC lighting, DC computers, etc. The PV station in Nam Co is a key demonstration base project of isolated grid PEDF built in an alpine and high-altitude environment, having typical demonstration significance. Built and put into use since October 2021, the station's system has been operating stably in sunny weather, and its stored energy has been able to support it to operate for more than one day in continuous cloudy and rainy weather. The ticketing personnel of the Holy Elephant Gate are stationed here to work in a comfortable working space and access energy.

(2) Zhuhai Hebao Island 5G Base Station

The system of the Hebao Island isolated grid PEDF 5G base station project (Fig. 7.8) fully utilizes PV green energy to minimize traditional fossil energy consumption, without polluting the environment. While ensuring cleanliness and high efficiency, it truly realizes isolated grid operation and provides a model for zero-carbon base station construction. The base station was built and put into use in March 2022. When the sunlight is abundant in the daytime, the base station operates in full reliance on PV power generation, and the excess power is charged to the energy storage unit; when there is no sunlight at night, the PV panels stop generating power, and at this time the energy storage unit seamlessly changes to discharging power to meet the uninterrupted power consumption of the base station. The stored backup power can meet load operation for 9.35 h in case of extreme continuous overcast and rainy weather.

7.4 Isolated Grid PEDF System Technology

Fig. 7.7 PV station in Holy Elephant Gate, Nam Co

(3) Off-grid Villa in Nigeria

Located in Lagos, Nigeria, the project is an Australian-style lightweight steel structure villa. Using dual power supplies from the local municipal power grid and a diesel generator, the original system had problems such as an unstable power supply, frequent power outages, loud noise, environmental pollution, and low efficiency. Based on the application environment in Lagos, Nigeria, the project uses the PEDF system that adopts a multi-energy complementary architecture involving PV, energy storage, municipal power, and diesel generator to build a 24 h uninterrupted power system, as shown in Fig. 7.9. In the conventional home scene, with a long power outage, the food in the refrigerator often rots. In this project, according to the four-level control method, the unit system regulates the load to achieve the stable operation of the isolated grid PEDF system. As a result, the refrigerator has never experienced power outages. The villa is currently operating stably under off-grid conditions, with the system meeting the daily power demand of a family.

Fig. 7.8 Zhuhai Hebao Island 5G base station

Fig. 7.9 Off-grid villa demonstration project in Nigeria

7.5 PEDF Power Storage-Swapping-Testing Integration Technology for Rural Areas

7.5.1 Proposing of the PEDF Storage-Swapping-Testing Integration Technology

The *Action Plan for Carbon Dioxide Peaking Before 2030* issued by the State Council proposes, "We will increase the proportion of buildings' energy consumption on electricity, and construct buildings integrating photovoltaic power generation, energy storage, DC power distribution, and flexible power consumption (PEDF)". Rural areas, as one of the main fields involved in the country's dual carbon goals, have great potential for developing PV resources. Promoting the PEDF technology in rural areas and building clean, safe, reliable, and economical rural energy infrastructure are guarantees of realizing rural revitalization and building a beautiful and harmonious countryside.

Rural areas have relatively weak power Infrastructure, where there are abundant rooftop resources and open tidal flats available for DG PV deployment, but the grid consumption capacity is poor. PV power fluctuations also greatly affect the safety, stability, and power supply quality of the grid. In addition, environmental protection, low-carbon, and people-benefiting actions, such as home appliances and electric vehicles going to the countryside, the coal-to-electricity project, and all-electric kitchen transformation, have increased the requirements for rural grid expansion and power supply guarantee, involving high investments in the capacity expansion and transformation of the traditional rural grid. Furthermore, the ongoing electrification of agricultural machinery and implements helps advance low-carbon production in rural areas but places a higher demand on the rural grid. Owing to the highly intermittent agricultural machinery operation and the low year-round utilization, long idle time, and insufficient professional maintenance of the power storage part that occupies the largest proportion of the cost of all-electric agricultural machinery, a more economical, safer, and more professional master plan is required.

In response to the above issues, a PEDF storage-swapping-testing integration technology for rural areas based on "storage and swapping standardization" is proposed based on the general idea of PEDF. At ordinary times, the standard storage modules are placed in the storage-battery swap cabinets for the stationary energy storage of the rural grid, for consuming PV power, balancing the peak-valley difference of the rural grid and improving the power quality, and for emergency power supply. During busy farming seasons, they are inserted into agricultural machinery and implements (those with swappable batteries have a lower cost) for use. In this way, the expensive storage modules are shared for multiple purposes, used to realize higher economic benefits, and are maintained and managed more professionally. In township transportation, standard power storage modules can also be used for small drainage and irrigation machinery, harvesters, small pickup trucks, etc., with single or multiple standard storage modules selected based on travel distance. Standard storage modules can also be used to supply power to field stations and pre-cooling cabinets during harvest seasons in a mobile way, facilitating use in multiple scenarios. To sum up, the PEDF storage-swapping-testing integration technology for rural areas is an economical, feasible, safe, reliable, flexible, and multi-purpose low-carbon energy solution.

7.5.2 Technical Principles of PEDF Storage-Swapping-Testing Integration System for Rural Areas

The PEDF system architecture is at the core of the system, as shown in Fig. 7.10. Taking the household solution as an example, the architecture can effectively realize flexible connection and integration of PV power generation, energy storage, and DC load through the integrated PV-storage-charging-swapping-testing cabinet, to create an economical, safe and reliable low-carbon energy system.

Based on the architecture, an integrated PV-storage-charging-swapping-testing cabinet, and a DC 48 V standard storage module have been developed, which store power from PV power generation or during valley periods, for discharging or battery swap use when needed. The integrated PV-storage-swapping-testing cabinet is used

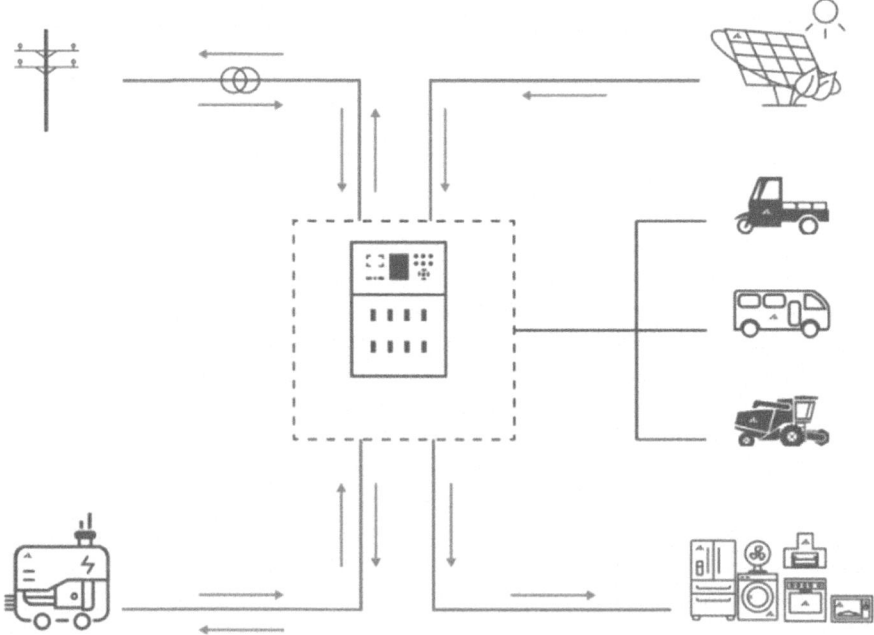

Fig. 7.10 Architecture of PEDF storage-swapping-testing integration system for rural areas

to connect PV power and intelligent electric apparatus, charge or swap the batteries of electric vehicles and agricultural machinery and implements, balance the supply–demand differences of the power grid, and connect methanol and other hydrogen fuel cells to improve the system's emergency backup and cross-cycle power supply guarantee capabilities. The DC busbar connects each unit to achieve the direct drive and direct use of "source, storage, grid, load, and backup", to fully tap DC advantages and improve system efficiency and stability. The PEDF information management system of integrated PV-storage-charging-swapping-testing cabinet realizes information monitoring, connectivity, and operation scheduling of each unit, and can provide interactive regulation of grid information to better coordinate the upper level of energy exchange and supply.

7.5.3 Technical Characteristics of PEDF Storage-Swapping-Testing Integration System for Rural Areas

Taking electric agricultural machinery and tools as an example, large machinery (such as tractors) is powered by multiple modules in parallel, while low-power machinery (such as short-distance vehicles) uses fewer modules for energy supply. The agricultural machinery and tools are seasonally and intermittently used. When agricultural machinery and electric vehicles are idle, the standard storage modules are installed back into the storage-swapping-testing integration system and become an indoor or village-level energy storage system. This can better consume rooftop PV power, improve the quality of the rural power grid, and support its steadier operation and emergency power supply.

As the core terminal device, the integrated PV-storage-charging-swapping-testing cabinet based on the "storage and swapping standardization" has a built-in multiport converter. It can be compatible with different modules and is a user-friendly interface for safe storage, professional management, and rapid swapping of DC 48 V standard storage modules. It supports parallel expansion and hot plug, and adapts to autonomous swapping of electric vehicles (bicycles, motorcycles, tricycles, engineering machinery, transport trucks, and passenger cars) to improve the efficiency of equipment resource utilization and reduce power costs.

What's more, the system has good scalability and can be flexibly expanded and upgraded according to different requirements. The developed DC 48 V standard storage module can be easily replaced and technically upgraded, and has more power level options and better parallel-series connection capabilities to meet the diversified needs of different users.

7.5.4 Application of PEDF Charging-Swapping-Testing Integration Technology for Rural Areas

(1) Application of Low-speed Electric Vehicles in Rural Areas

The 48 V low-speed electric vehicle with the PEDF charging-swapping-testing integration system may be available for rural and inter-rural commuting traffic, as shown in Fig. 7.11. Rural roads are characterized by short mileage and complicated road conditions, and the vehicles are expected to have excellent wading capacity and strong passability. Low-speed electric vehicles of 100 km/h and below are important options. The 48 V scheme is known for low voltage, low insulation requirements, and fast breakthrough in the LV parallel drive technology. The low-speed electric sedan and electric pickup truck developed by this scheme are more economical and reliable.

(2) All-electric Applications of Agricultural Machinery and Tools

The integrated PV-storage-charging-swapping-testing cabinet for rural areas can be used in rural facilities such as irrigation systems, solar supplement systems, and electric agricultural machinery (electric mini-tillers and electric tractors).

(3) Application of capacity expansion of rural power grid and protection of transformer operation in the transformer coverage area.

The eco-friendly low-carbon promotion efforts such as the popularization of electrical appliances and electric vehicles in rural areas, "coal-to-electricity" and electrified kitchens have increased the demand for rural power grid expansion and electricity retention. The PEDF technology and rural storage-swapping-testing integration system are better approaches to achieving clean, low-carbon, safe, and reliable

Fig. 7.11 Application scenario of low-speed electric vehicles in rural areas

Fig. 7.12 Application scenario of integrated PV-storage-charging-swapping-testing cabinet in the transformer coverage area

development of the rural energy system in the long term, improving the quality and efficiency of the livelihood of rural households and developing rural industries (Fig. 7.12).

In conclusion, the PEDF technologies underlain by "storage-swapping standardization" such as the DC 48 V standard storage module, low-speed e-mobility and integrated PV-storage-charging-swapping-testing cabinet and the storage-swapping-testing integration technology, systems and equipment for rural areas can be the supporting technical route for rural revitalization. Application of these technologies in various scenarios has proven that they are suitable for rural areas, and they are proven to be economically feasible, safe, reliable, and reliable multi-purpose low-carbon energy solutions.

7.5.5 Pilot Application and Development of Rural PV-Storage-Swapping-Testing Integration Technology

Now domestic research institutes and new ventures are conducting research and development in this regard. The Guochuang Energy Internet Innovation Center (Guangdong) Co., Ltd. has conducted research into the PEDF storage-swapping-testing integrated energy station system. Chongqing University and Chongqing Baizhuan Technology Co., Ltd. have developed the 48 V "storage-swapping-testing" integrated agricultural machinery and the three electrical systems for rural town traffic, and have developed the electric agricultural machinery, electric logistic vehicle, agricultural electric pump, electric mini-tiller and electric express delivery motorcycle that support storage-swapping.

Fig. 7.13 PV future roof DC community DC 48 V storage-swapping research pilot project

(1) PV Future Roof DC Community DC 48 V Storage-Swapping Research Pilot Project (Fig. 7.13)

The Guochuang Energy Internet Innovation Center (Guangdong) Co., Ltd. has developed the PEDF swapping-testing demonstration cabinet and swappable cart in its PEDF DC community demonstration base. The main components of the demonstration cabinet include a PV module, 48 V battery pack, DC cavity, energy storage system, and DC lighting system. The DC generated by the PV array whose surface receives solar radiation is transmitted as DC 400 V electricity to the cavity and energy storage system via the hybrid power system. The output AC 220 V current is diverted to the mains supply system. The DC 400 V bus is connected to a DC/DC power converter that outputs DC 48 V current for interior lighting, projector, and other LV DC loads. The 48 V standard battery packs can be used by swappable carts, pickup trucks, sedans, and logistic vehicles, allowing people to travel to remote locations such as suburban areas. This pilot project has demonstrated the feasibility of the DC system, meets the energy storage and battery swap requirements, and supports both fixed and mobile applications.

(2) Dazu Tianhua Baihui Park PV-Storage-Swapping-Testing Integration Technology Demonstration

The Dazu Tianhua Baihui Park Project of Chongqing occupies a land area of 2000 mu (1.33 km^2), but another 2000 mu (1.33 km^2) will be added. The seasonal loads require 30 kW of power now. The transformer is installed at a substation 3 km away with a dedicated power line and its capacity is 400 kVA. The load in spring is thousands of kilowatts, so the transformer is virtually in a non-load state, and this entails high no-load loss and high construction cost. There are also problems including high construction costs, low utilization rate of equipment, and irrational load allocation. The other power supply line is a three-phase four-wire LV line from a transformer coverage area 1 km away and is routed to the greenhouse area, with a low operating load. The park has 70–80 workers, and they need to commute by using two-wheel electric vehicles and small three-wheel electric vehicles. The park has three-wheel electric vehicles that are used to transport flowers and tools.

7.5 PEDF Power Storage-Swapping-Testing Integration Technology … 233

Fig. 7.14 Rural electrification pilot project of State Grid—Chongqing Baihui Park

In the Chongqing Baihui Park Project (Fig. 7.14), the 2022 rural electrification pilot project of State Grid Chongqing Dazu Power Supply Co. Ltd., is intended to enhance the power supply capacity of the rural areas, increase the income of the rural households and support agricultural development through the efforts such as upgrading of rural power grid, improvement of power supply services, intensified industrial assistance and consumer assistance.

The park contains the 48 V distributed integrated PV-storage-charging-swapping-testing cabinet that supports swappable carts, pickup trucks, sedans, and logistic vehicles, and provides a 15 kW power supply. A customized 48 V electric system that consists of a power generator, electric controller, and battery has been developed for the project. Through the above measures, this park provides the energy storage and emergency power supply used for plant cultivation experimentation equipment of the flower research institute, manages to inhibit the voltage fluctuation of the power grid, solves the problems of load fluctuation and low voltage of the greenhouses, provides the autonomous battery swapping service for carts, tricycles, logistic vehicles, and passenger cars, and provides the driving power for vehicles, travel tools, in-park electric machinery and agricultural machinery. This project demonstrates the advanced nature of PEDF technology, the reliability of PV-storage-swapping-testing integration technology, enhances the level of electrification of the rural areas, increases the utilization rate of energy assets, and significantly improves the development level of the rural areas.

7.6 Tracking-Free Solar Heat Collection Technology to Meet the Seasonal Heating Requirements for RRB

7.6.1 Technical Limitations of Traditional Solar Water Heating Systems

Due to the instability of solar energy resources, the seasonal fluctuation trend of solar energy goes contrary to the fluctuation trend of user-side heating requirements. In winter, the solar point moves to the Southern Hemisphere, resulting in a decrease in solar irradiation intensity and a shorter sunshine duration. Allowing for the rising demand for space heating and domestic hot water, the system can make up for the shortage of solar heating by increasing the collector area or the input in auxiliary heating equipment. In summer, there is no demand for heat supply and low demand for hot water, but summer has high irradiation intensity and long sunshine duration, so in the case of clear weather, the water tank will reach full load overly soon and the circulating water pump will stop working. When the system is in stagnation state, the collector exposed to sunshine is overheated very quickly. The high-temperature and high-pressure steam generated will increase the risk of damage to the heat collector and system modules, as shown in Fig. 7.15. Besides, high temperature accelerates the material aging, and will lead to degradation of antifreeze when antifreeze is used as a circulating working medium. In the "supply less than demand" situation in winter, the stability and economic efficiency of the solar heating system are reduced. In the "supply more than demand" situation in summer, the safety and reliability of the solar heating system are reduced. This is why the solar heating system is faced with the problem of "water not hot enough in winter and too hot in summer".

Fig. 7.15 System problems caused by excessive heat in summer

7.6.2 Principle of Heat Collection Technology Based on Seasonal Adaptability of Heating Requirements

The thermal output of traditional solar collectors varies passively with the solar energy fluctuations. The fundamental solution is to keep the heat output of the collector adapting to the seasonal changes of heating requirements, i.e., increase in winter and decrease in summer. Given the seasonal variation of the subsolar point, the elevation angle of radiation is low in winter and high in summer. Figure 7.16 shows the angle variation of the effective direct solar radiation in different seasons in Beijing. It can be clearly seen that the effective direct solar radiation in winter and summer forms a clear interval separation of the angle of incidence.

Given the seasonal variation of the elevation angle of solar radiation, directed solar radiation concentration in winter and solar radiation scattering in summer are executed by using the angular constraint function of a non-imaging concentrator. In this way, the collector can switch between high photo-thermal performance in winter and low photo-thermal performance in summer without using any mechanical device. The 2D profilogram of a small non-imaging concentrator is shown in Fig. 7.17. The basic function of this reflective board is to focus the incident ray with an elevation angle less than θ_C onto the absorber, and reflect the incident ray with an elevation angle more than θ_C back into the sky, so as to adjust the optical performance of the equipment seasonally according to the solar elevation angle. As shown in Fig. 7.17a, if the angle of incidence is less than the critical acceptance angle, the reflected ray parallel to the incident ray is focused on the underside of the heat-absorbing tube. Since a non-imaging concentrator is used here, the reflected ray will have no fixed focal point. If the incident elevation angle is greater than the critical acceptance angle,

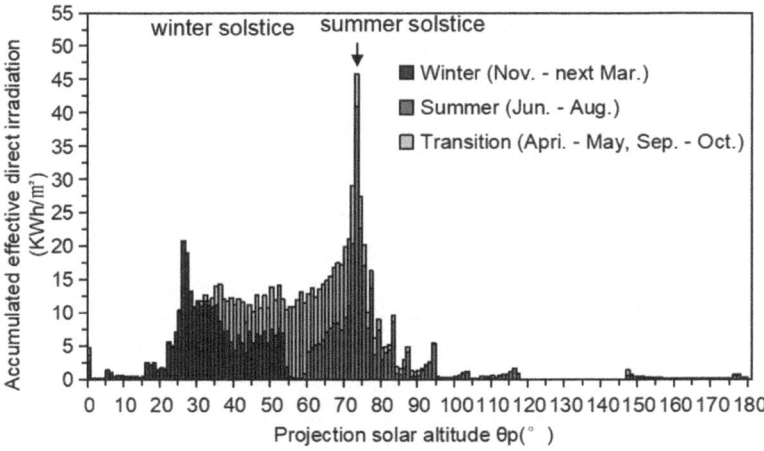

Fig. 7.16 Characteristic difference of solar radiation in winter and summer [A case study of Beijing (37.45°N)]

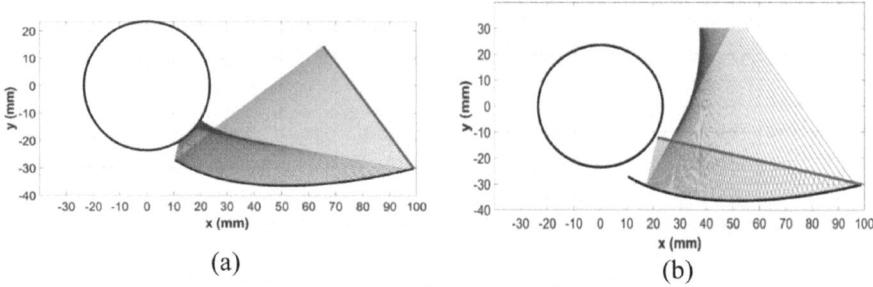

Fig. 7.17 Schematic diagram of an angle-constrained concentrator

the reflected incident ray will not be focused on the heat-absorbing tube and instead will be all reflected back to the sky, as shown in Fig. 7.17b.

7.6.3 Technical Assembly Modes in Different Application Scenarios

Figure 7.18 shows the structure of the new-type collector. The main structure of the prototype is mainly composed of 20 U-shaped vacuum tubes with a length of 1.8 m and concentrator blades arranged horizontally. The transverse vacuum tube and reflector blades are fixed by the left and right side frame supports. 20 vacuum tubes are connected in parallel. The U-shaped copper tube inside each vacuum tube is connected to the manifold in the right sealing header. During operation, the low-temperature water flows into the inlet manifold and then diverts to 20 parallel U-shaped pipe branches. The high-temperature hot water after heat exchange flows into the return manifold and then out of the collector. Glass insulation cotton is inserted into the header to insulate the manifold. To locate the reflector, a small triangular support is installed at the end position of each blade reflector with bolts at the frame support rails on both sides, and then the reflector is fixed on the triangular support at both ends.

Because the designed concentrator blade can flexibly rotate angle and expand width as required, and the angle of inclination and tube pitch of the whole row of vacuum tube absorbers can be adjusted flexibly, different assembly modes can be realized through the free combination of parameters, thus adapting to different application scenarios. As shown in Fig. 7.19, the new low-temperature concentrating collector can be installed independently or integrally on sloping roofs or vertical walls depending on the scenarios.

7.6 Tracking-Free Solar Heat Collection Technology to Meet the Seasonal … 237

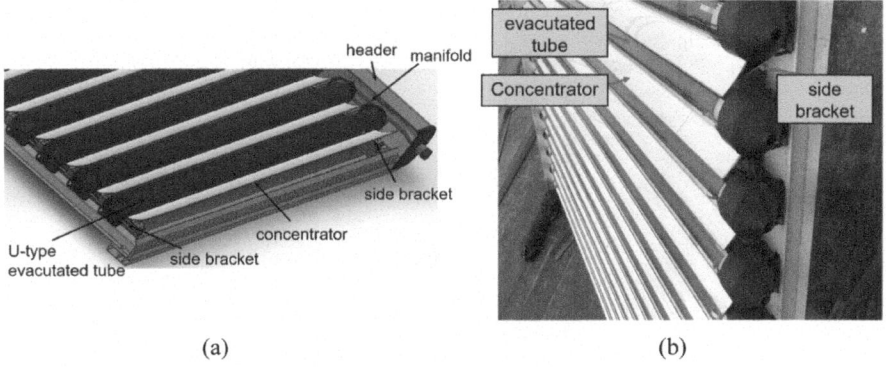

Fig. 7.18 Schematic diagram of the new-type collector

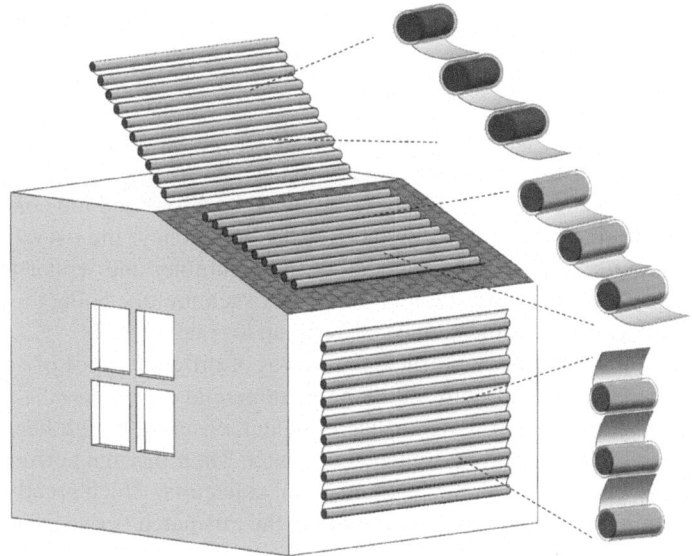

Fig. 7.19 Assembly modes in different application scenarios

7.6.4 Plant Performance Tests

To verify the seasonal performance of new-type collector, the collectors with coupled concentrators and the vacuum tube collectors without coupled concentrators are tested outdoors for a long period to analyze their operating performance changes under the same conditions. Figure 7.20 shows the optical efficiency in different seasons. It can be seen that due to the functional switch of the reflector in winter and summer, the average daily optical efficiency of new-type collectors is significantly

Fig. 7.20 Comparison of optical efficiency tests in winter and summer

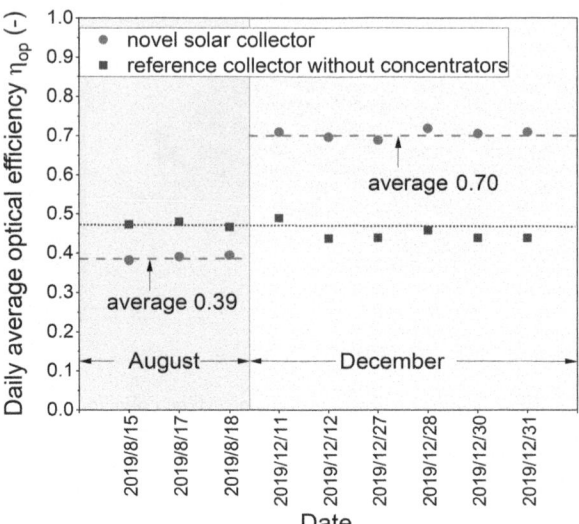

different in summer and winter. In August, the solar elevation angle is high, and the concentrator is in the defocusing state, giving rise to low optical efficiency, averaging 0.39. In December, the solar elevation angle is low, and the concentrator is in the concentrating state, producing high optical efficiency; the overall average daily optical efficiency is 0.70. In contrast, on both summer and winter test days, the average daily optical efficiency of conventional vacuum tube collectors without coupled reflectors remains almost constant, with an average of 0.46.

Figure 7.21 shows the thermal efficiency curves of different types of collectors. It can be seen that compared with traditional commercial collectors, the new-type collectors have better thermal performance in winter, ensuring a high heat production capacity for high heating requirements in winter. Their thermal performance in summer is lower than all conventional commercial collectors, which greatly reduces the risk of system overheating in summer due to the mismatch between supply and demand.

Figure 7.22 compares the heat collection performance of traditional flat plate collectors and new-type collectors in winter and summer under different urban application scenarios. For detailed data, see Table 7.2.

Compared with the traditional flat plate collector, the new-type collector features significantly higher heat collection in winter, thanks to the high performance of the track-free miniature concentrator. This is very beneficial for high-latitude severe cold areas. As shown in Fig. 7.22a, compared with the traditional equipment, the cumulative heat collection per unit area of new-type collectors in Harbin and Hohhot in winter increased by 56% and 40%, respectively. Meantime, the new-type collector reduces the thermal performance by defocusing in summer, so as to avoid the high-temperature overpressure damage caused by greater supply than demand in summer.

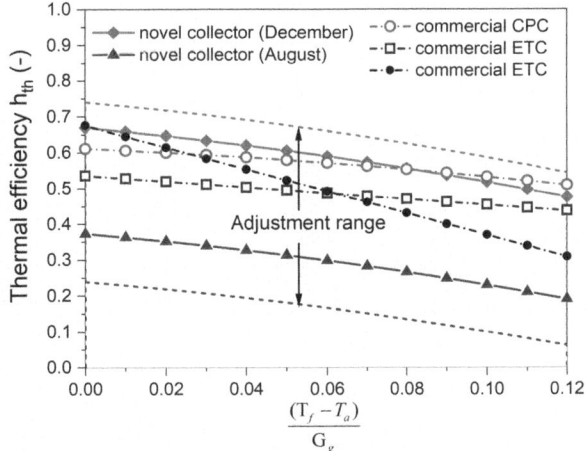

Fig. 7.21 Transverse comparison of performance between new-type collectors and traditional collectors

As shown in Fig. 7.22b, compared with traditional collectors, the new-type collector achieves a heat reduction ranging from 29 to 32% in cities.

7.7 Summary

(1) This technology breaks the passive variation of the thermal output of traditional solar collectors with the seasonal change of solar radiation, and realizes the output of increase in winter and decrease in summer through the angle-constrained condensing mechanism, so as to overcome the technical bottleneck of the source-load mismatch from the system source side.
(2) The new collector configuration is flexible and extensible, and can form flexible assembly modes for different application scenarios to adapt to independent installation, or integrated installation on sloping roofs and vertical walls.
(3) Compared with other traditional commercial collectors, the new collector achieves higher performance in winter and lower performance in summer through the track-free form, which better matches the characteristics of seasonal changes in the user's heat demand. While ensuring high-performance output in winter with high heating requirements, it reduces the risk of system overheating in summer.

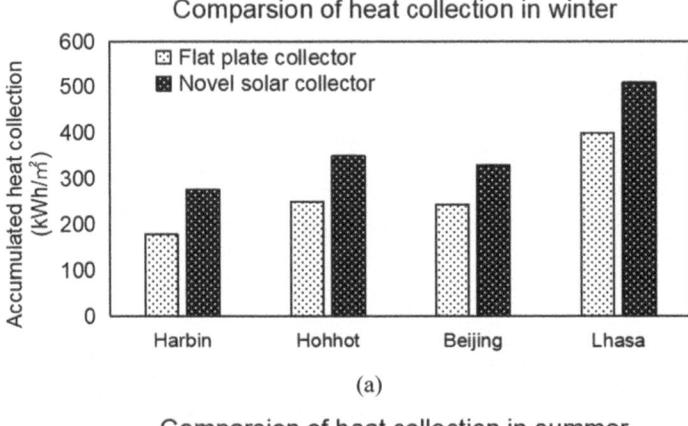
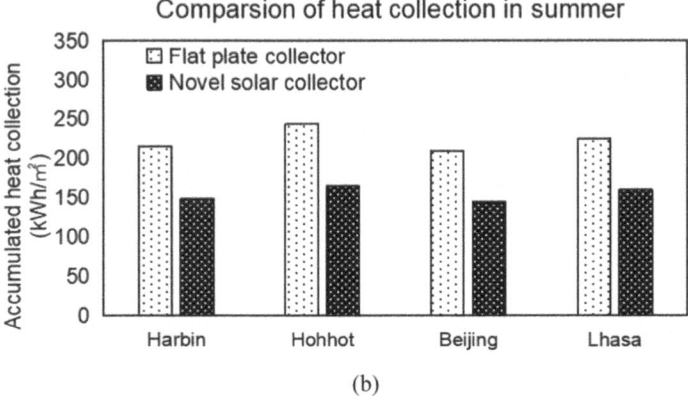

Fig. 7.22 Application effects of new-type collectors vs. traditional flat plate collectors in different cities. **a** Collectors in winter; **b** Collectors in summer

7.8 Biomass Pyrolysis and Gasification Technology

7.8.1 Basic Principles of Pyrolysis and Gasification Technology

Biomass gasification is the process of converting hydrocarbons that makeup biomass into combustible gases, such as CO, H_2, and low-molecular hydrocarbon gases (such as CH_4), under certain thermal conditions.

The equipment used for biomass gasification is a biomass gasifier. It comes in many types. Now the working process of a down-flow (aka down-draft) fixed bed gasifier is taken as an example to illustrate the basic principle of biomass gasification (Fig. 7.23).

First, part of the gasification feedstock is added from the top of the fixed bed gasifier, dropped onto the grate bar, and lit with a tinder. Then the feedstock is

7.8 Biomass Pyrolysis and Gasification Technology

Table 7.2 Comparison of winter and summer data in different cities

City	Traditional flat plate collector		New-type collector		Heat increase in winter (%)	Heat reduction in summer (%)
	kWh/m^2) Winter (kWh/m^2)	kWh/m^2) Summer (kWh/m^2)	kWh/m^2) Winter (kWh/m^2)	kWh/m^2) Summer (kWh/m^2)		
Harbin (45.7N°, 126.6E°)	177.0	214.4	275.4	148.0	56	31
Hohhot (40.8N°, 111.7E°)	249.9	243.4	349.1	164.5	40	32
Beijing (39.9N°, 116.4E°)	243.7	208.5	328.9	144.3	35	31
Lhasa (29.7N°, 91.1E°)	398.5	224.8	509.4	159.5	28	29

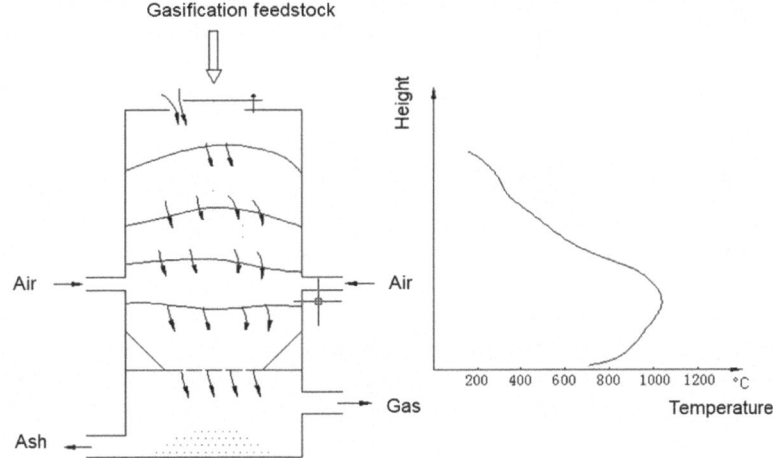

Fig. 7.23 Schematic diagram of biomass gasification principle

gradually added to the required height. Turn on the blower, and the air is drawn into the gasifier through a number of air inlet nozzles around the gasifier. The upper and lower positions of the nozzles are generally located near the "oxidation zone" to allow some air to be sucked into the gasifier from the feed port on the top of the gasifier. During the working process, the airflow in the gasifier flows from top to bottom and under slightly negative pressure conditions. The resulting combustible gas (gas for short) passes through the grate bar and is extracted through the air outlet.

The gasification feedstock falls into the gasifier from the feed port on the top of the gasifier, either continuous feeding or intermittent feeding.

When the biomass gasifier is started to function properly, the feedstock above is continuously added, and the materials in the gasifier are slowly moving downward. The whole gasification process can be roughly divided into four zones from top to bottom: drying zone, thermal decomposition zone, oxidation zone, and reduction zone. The reaction process in each zone is as follows:

(1) Drying Zone—Evaporation of Water

The feedstock containing water is exchanged with the underlying heat source here, and the temperature rises to 100–300 °C to evaporate water in the feedstock and become dry materials. Water vapor flows downward under the action of suction, and dry materials move downward under the action of gravity.

(2) Thermal Decomposition Zone—Devolatilization

Materials, water vapor, and air from the drying zone continue to obtain the heat transferred from the oxidation zone after entering the thermal decomposition zone. When the temperature of the material rises to a certain value (minimum 160 °C), the biomass will undergo pyrolytic reaction and precipitate volatiles. The temperature in the thermal decomposition zone is 300–700 °C. Thermal decomposition reaction products are somewhat complex: main volatiles are H_2, water vapor, CO, CO_2, CH_4, tar, and other hydrocarbons, and the remaining solid material for coke (also known as charcoal).

(3) Oxidation Zone—Oxidation Reaction

The product in the biomass thermal decomposition zone keeps moving downward together with the air and water vapor in the gasifier, so the temperature will rise again. When the temperature reaches the minimum fire point of the thermal decomposition gas (250–300 °C), the combustible volatile component will be ignited and combusted. Then the blazing charcoal experiences imperfect combustion, to generate CO, CO_2, and water vapor. The oxidation zone has a fast reaction rate to release large amounts of heat, where the maximum temperature can be 1000–1200 °C. The volatile component will be combusted and further degraded. It is the reaction heat generated in this zone that provides the heat source for the drying zone, the thermal decomposition zone, and the reduction zone.

The main chemical reactions in the oxidation zone are

$$C + O_2 \rightarrow CO_2 \qquad (7.6.1)$$

$$2C + O_2 \rightarrow 2CO \qquad (7.6.2)$$

$$2CO + O_2 \rightarrow 2CO_2 \qquad (7.6.3)$$

7.8 Biomass Pyrolysis and Gasification Technology

$$2H_2 + O_2 \rightarrow 2H_2O \tag{7.6.4}$$

$$CH_4 + 2O_2 \rightarrow CO_2 + 2H_2O \tag{7.6.5}$$

(4) Reduction Zone—Reduction Reaction

There is almost no O_2 in the reduction zone, and CO_2 and high-temperature water vapor react with hot carbon here to produce CO and H_2. These gases and volatiles form combustible gases and complete the conversion process of solid biomass to gas fuel. Since the reduction reaction is endothermic, the temperature of the reduction zone is also reduced to 700–900 °C, with a slow reaction rate. The main chemical reactions in the reduction zone are:

$$C + CO_2 \rightarrow 2CO \tag{7.6.6}$$

$$H_2O + C \rightarrow CO + H_2 \tag{7.6.7}$$

$$C + 2H_2 \rightarrow CH_4 \tag{7.6.8}$$

$$H_2O + CO \rightarrow CO_2 + H_2 \tag{7.6.9}$$

$$3H_2 + CO \rightarrow CH_4 + H_2O \tag{7.6.10}$$

In the gasification process, the gas in the gasifier goes down through the fidley and is pumped from the gas outlet by the blower. As the feedstock is constantly added, the materials in the gasifier are successively moved down, turned into ash, welded to the ash chamber, and pulled out from the ash outlet.

The oxidation zone and reduction zone are collectively called the gasification zone, where the gasification reaction takes place. The drying zone and the thermal decomposition zone are collectively called the combustion preparation zone. It should be noted that the distinct division of the gasification process into several zones is not exactly in line with the actual situation, merely for the sake of analysis. In fact, one zone may be partially incorporated into another. As a result, these processes may intersect.

At present, most biomass gasification stations in China employ the down-flow fixed bed gasifier for gasification. Table 7.3 shows the main components and calorific value of produced gas from some manufacturers.

Table 7.3 The main composition of gas and lower calorific value

Raw material	Gas compositions (%)						Lower calorific value under standard conditions (kJ/m^3)
	CO	H$_2$	CH$_4$	CO$_2$	O$_2$	N$_2$	
Cornstalk	21.4	12.2	1.87	13.0	1.65	49.88	5328
Corncob	22.5	12.3	2.32	12.5	1.4	48.98	5033
Wheat straw	17.6	8.5	1.36	14.0	1.7	56.84	3663
Cotton stalk	22.7	11.5	1.92	11.6	1.5	50.78	5585
Rice husk	19.1	5.5	4.3	7.5	3.0	60.5	4594
Firewood	20.0	12.0	2.0	11.0	0.2	54.5	4728
Leaves	15.1	15.1	0.8	13.1	0.6	54.6	3694
Sawdust	20.2	6.1	4.9	9.9	2.0	56.3	4544

Note Standard conditions refer to 0 °C and 1 physical atmospheric pressure

7.8.2 Product Performance

Based on biomass resource characteristics and products, different biomass gasification equipment and process routes are available: The fluidized bed polygeneration gasifier is used for crop straw raw materials, the fixed-bed down-draft polygeneration gasifier for raw materials of fruit shells, and the fixed-bed up-draft polygeneration gasifier for wood raw materials. Appropriate product utilization routes can apply to different biomass gasification products: The gaseous products (biomass gas) are used for power generation, gas supply, direct-fired boiler heating, or driving steam turbines to generate electricity; solid phase product (biomass charcoal) can be used to prepare carbon-based organic–inorganic compound fertilizers (straw raw materials), high value-added activated carbon (fruit shells and wood chips), and industrial reductant and civil fuels (wood-based) depending on the biomass feedstock. Biomass gasification polygeneration technology has explored a green, environmentally friendly, circular, and sustainable development path for biomass utilization.

(1) Biomass Gas Performance

Biomass gas mainly contains combustible components such as CH$_4$, CO, and H$_2$, in addition to non-combustible components such as N$_2$, CO$_2,$ and a small amount of O$_2$. The gas composition and calorific value of different raw materials are shown in Table 7.4.

(2) Biomass Charcoal Properties

Depending on the source of raw materials, biomass charcoal can be divided into rice husk charcoal, charcoal, straw charcoal, etc. Table 7.5 shows the related parameters of rice husk and wood chip.

7.8 Biomass Pyrolysis and Gasification Technology

Table 7.4 Gaseous compositions and calorific values of different raw materials

Raw material	CH_4 (%)	CO_2 (%)	CO (%)	H_2 (%)	O_2 (%)	Calorific value (kJ/m^3)
Rice husk	2.60	13.50	14.50	9.50	1.61	3762
Wood chip	3.94	19.62	12.62	12.35	0.27	5300
Apricot shell	3.32	17.64	10.84	14.87	0.27	4351

Table 7.5 Parameters of rice husk and wood chip

Raw material	Calorific value (kJ/kg)	Ash content (%)	Volatile content (%)	Fixed carbon content (%)
Rice husk	18,497	45.35	5.21	49.44
Wood chip	30,188	8.44	9.76	81.80

Biomass charcoal is a porous carbon material, black, mainly powdery and granular. Biomass charcoal features developed pore structure, large specific surface area, specific surface functional groups, and stable physical and chemical properties. It is acid and alkaline-resistant, can withstand moisture, high temperature, and high pressure, insoluble in water and organic solvents. In particular, it can be made into activated carbon, which is an excellent adsorption and purification material and is also used as a catalyst or catalyst support. It is an indispensable and important material for industry, agriculture, national defense, transportation, medicine and health, environmental protection, and cutting-edge science.

The charcoal-based compound fertilizer (charcoal content 20–30%) produced from biomass charcoal has the following effects when used in farming: increase the soil porosity, reduce soil density, improve air and water permeability of soil, and increase the maximum water-holding capacity of soil; increase the air permeability of soil, mitigate the problem of soil hardening, and eliminate the effect of soil hardening; return the elements in short supply in the soil of China, such as nitrogen, phosphorus, potassium and magnesium, to the soil, and supplement the elements required by plants, such as copper, iron, and zinc, which will increase the yield and quality of the crops; inhibit phosphorus adsorption by the soil, improve desorption of phosphorus, and thus improve phosphorus adsorption and ingestion by plants; reverse the heavy metal contamination of soil, and significantly desorb the cadmium (Cd) in the contaminated soil; increase the temperature of soil (1–3 °C), improve growth of crops, and shorten the growth and maturation period (by about 7 days); stabilize the temperature of soil (1–3 °C), improve growth of crops, and shorten the growth and maturation period (by about 7 days); stabilize the pH of soil, slow the release of fertilizer and pesticide, enhance water retention of soil, improve the microbial environment of soil, and increase the lodging resistance of rice; biomass charcoal (carbon content 50–90%) can also improve CO_2 fixation, where 1t of biomass charcoal can fixate over 2t of CO_2.

Table 7.6 Properties of the fluids extracted from rice husk

Sample	Density (g/cm^3)	Refractive index n_D^{19}	pH	Total acid (%)	Color	Smell
Crude extract	1.050	1.7856	6.86	0.89	Black brown	Burnt smell
Refined extract	1.025	1.3387	6.93	0.56	Yellowish, transparent	Slight burnt smell

(3) Composition and Properties of Biomass Extracted Fluids

The fluid extracted from rice husk comprises about 18 substances, including 21.17% alcohols, 3.75% phenols, 3.65% esters, and 40.08% acids, esters, aldehydes, and ketones. The properties of the fluids extracted from rice husk are listed in Table 7.6.

The fluid extracted from rice husk can eliminate the odor of chicken manure in a short amount of time, while phcnoium compositum cannot do so. The toxicology test on biomass extracted fluids indicates that the oral toxicity test of the specimen on male and female mice is non-toxic. The measurement of the heavy metals and bacteria in the biomass extracted active fluids indicates the content of heavy metals and bacteria in the specimen meets the standard *Skin Tonic* (QB/T2660-2004).

7.9 Integration of Natural Gas Production from Straws and Distributed Gas Supply

In biogas-based natural gas production, the raw materials are mainly straws, but are also mixed with livestock excrement and other organic wastes. The straws experience the fast chemical pretreatment process, and then experience anaerobic digestion with diverse mixed materials and anaerobic fermentation technology. The biogas generated is then purified by the pressurized washing and purification process. The biogas residue is used to produce organic fertilizer. A part of the biogas slurry is used in the anaerobic fermentation system, and the other part is used as liquid organic fertilizer to help crop growth. The entire system is controlled remotely and automatically.

7.9.1 Technical Principle

In this technology, straws are used as the main raw material. The material experiences fast chemical pretreatment, and is then mixed with the other materials such as agricultural organic waste, livestock excrement, and urban organic domestic waste. The mixture is put in an anaerobic fermenter for anaerobic fermentation. The biogas generated is purified to become the natural gas that is used for domestic and industrial purposes. The biogas slurry and biogas residue generated are separated from the liquid. The biogas residue is made into organic compound fertilizer, and the biogas slurry into leaf fertilizer. The CO_2 separated from the purification process can be used to produce industrial and food-grade CO_2. This process consists of six parts, including raw material collection, storage and transportation, fast chemical pretreatment, anaerobic fermentation with diverse mixed materials, environment-friendly biogas purification, comprehensive utilization of biogas slurry and biogas residue, and remote online automatic control. The technical route is shown in Fig. 7.24.

1. Raw Material Collection, Storage, and Transportation

Straws are quite scattered and their harvesting is highly seasonal, so the collection, storage, and transportation will be the main hindrance to their large-scale utilization. The existing distributed and intensive straw collection, storage, and transportation model comes short of competitiveness due to the need to maximize the profits, because the straw collection and transportation costs are too high, while a centralized urban gas supply system cannot afford such high costs. Therefore, the scope of raw material costs can be determined according to the quantity of raw materials required by the new urban centralized gas supply system and the quantity of

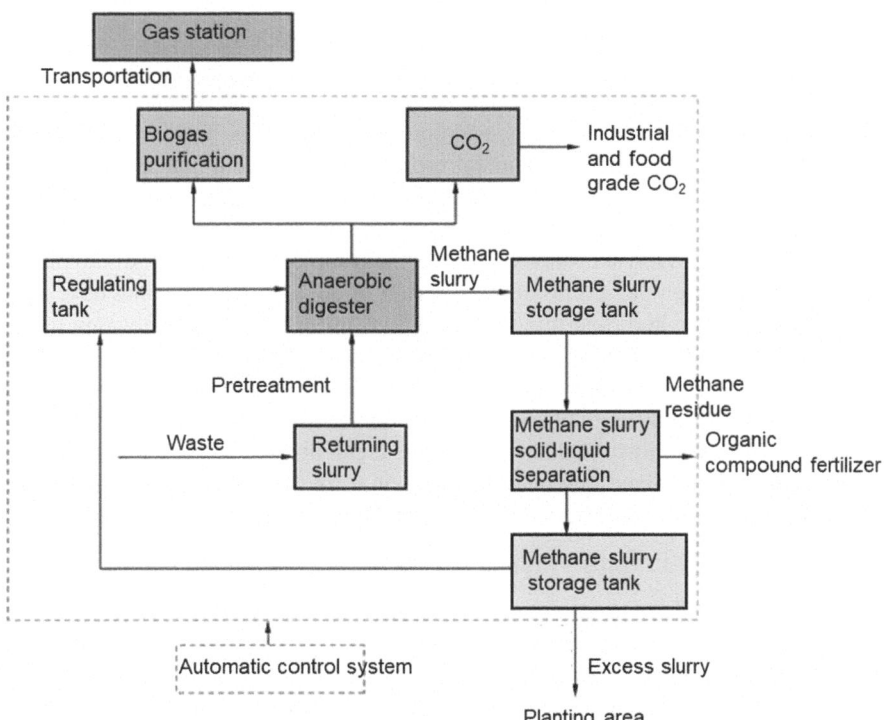

Fig. 7.24 Technical route diagram for biogas-based natural gas production from straws

supplied gas. The quantity of collected and transported materials and the collection and transportation costs can be controlled through the approaches such as "agricultural machinery replacement", "farm nanny" and "products replacement". These measures will increase the collection and transportation efficiency, reduce the costs, and help determine the optimum transportation distance under the given collected quantity and the collection and transportation costs.

2. Low-Cost Fast Pretreatment Technology

With the solid-state chemical pretreatment process at atmospheric temperature, the gas production from straws is increased by 50–120%, thus the gas production rate per unit of dry matter of straws far exceeds that of cow dung. The treatment process is explained by using corn stalks as an example. A special-purpose kneading machine kneads the corn stalk to destroy its physical structure and to make it easier to conduct immersion of chemical agents and to chemically process the lignocellulose in the corn stalk. The kneaded corn stalk is mixed with a certain quantity of special-purpose chemical agents, and the mixture is put in the pretreatment tank. Immersion in the chemical agent and chemical processing of the lignocellulose in the corn stalk will break the intrinsic bond between the lignin and cellulose/hemicellulose, change the crystallinity of cellulose, increase the contact area between corn stalk and anaerobe, and thus enhance the bio-digestibility of corn stalk and the gas production rate. The material can be taken out after it is kept at atmospheric temperature for 3 days. Then it is put in the anaerobic fermenter for anaerobic fermentation.

3. Near-Synchronous Synergistic Fermentation of Diverse Materials

The diverse materials such as straws, excrement, and domestic waste differ considerably in terms of physical and chemical properties and also in material properties and composition. For example, straws have a high content of C, livestock excrement has a high content of N, and domestic waste has a high organic matter content. It is necessary to achieve synchronous or near-synchronous fermentation of the mixed raw materials, shorten the gas production period of the raw materials that have long fermentation period, and determine the optimum mix ratio of different raw materials in anaerobic digestion, so as to ensure different raw materials can maximize their advantages under virtually same conditions. There is also the need to achieve directed acidification through phase separation, determine and regulate the C/N and C/P ratios, determine how the addition of trace elements will affect the gas production performance of mixed raw materials, and analyze the mechanism of achieving synergistic action through artificially adding and controlling trace elements.

4. Environmentally Friendly Washing and Purification Technology

Given the fact that different components of biogas have different levels of solubility in water, the CO_2 and H_2S can be removed by the pressurized washing process. The pressurized washing process consists of three subsystems, including the CO_2 and S removal system, condensation and dehydration system, and water regeneration system, as shown in Fig. 7.25. The feed biogas under atmospheric pressure is pressurized by the booster pump, and then enters the feed gas surge drum under a certain temperature. The gas with a certain constant pressure enters the absorber via the bottom of the absorber, while water enters via the top to achieve reverse flow adsorption, so as to remove the CO_2 and H_2S. The purified gas is discharged from the top of the absorber. Then the condensation and dehydration system removes the free water from the gas. Eventually, the acceptable product gas is obtained. The rich liquid is discharged from the bottom of the absorber, and then enters the regenerator for the purpose of regeneration by depressurization or air purging. The regenerated water enters the absorber again, thus completing the cycle. The pressurized water washing process is a green eco-friendly technology. The water used as absorbent can be used cyclically, and emission is zero, so no secondary environmental pollution happens. Besides, CO_2 and H_2S can be absorbed by water, which minimizes methane loss, increases the purified concentration, and reduces the investment and operation cost. In addition, the water is noncorrosive, thus reducing the equipment cost.

5. Comprehensive Utilization of Biogas Slurry and Biogas Residue

The biogas residue and biogas slurry discharged from the fermenter experience solid–liquid separation. The biogas residue is converted into sellable compound organic fertilizer. 50.9% of the biogas slurry is recycled and then used as feed regulating water. 20% of the biogas slurry is discharged and stored in the biogas slurry tank and is then directly used as liquid fertilizer. Biogas slurry is directly used as liquid organic fertilizer in vegetable greenhouses. The surplus biogas slurry is used for soil improvement in the cropland around.

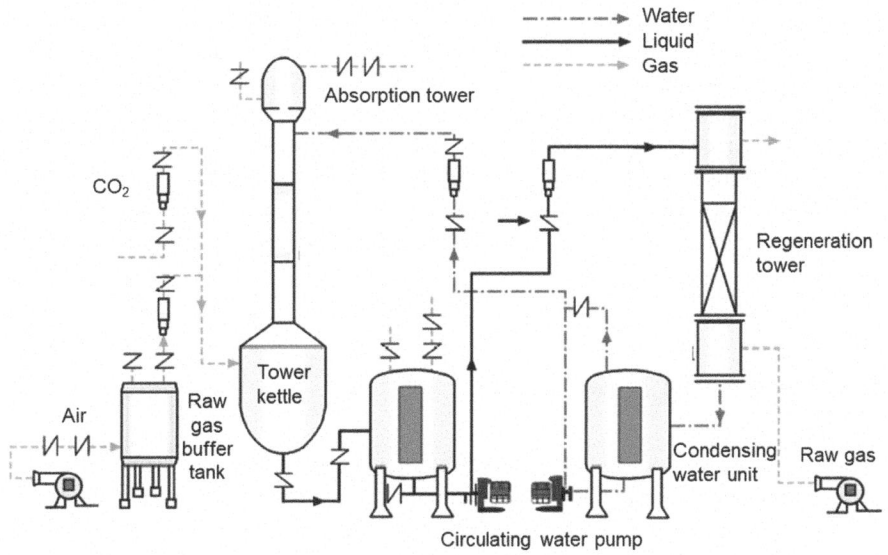

Fig. 7.25 Process flow diagram for pressurized water washing

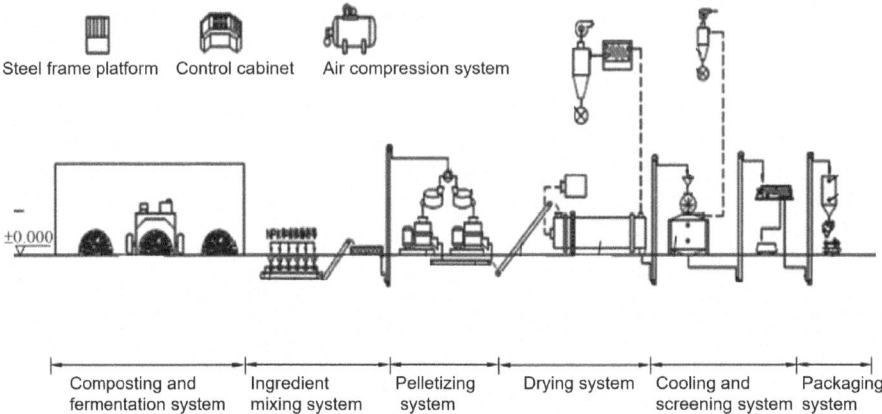

Fig. 7.26 Granular organic compound fertilizer production system

The granular organic and inorganic compound fertilizer production system comprises the compost fermentation system, materials mixing system, granulation system, drying system, cooling and screening system, packaging system, and control system, as shown in Fig. 7.26.

6. Remote Online Automatic Control

The automatic control system of a biogas-based natural gas project is meant to monitor, control, interlock, and alert the process parameters, electrical parameters, and equipment operating state of the straw anaerobic fermentation system and biogas purification system and also to print reports. A set of communication links is used to fulfill the requisite functions of the entire process such as data acquisition, data communication, sequential control, time control, loop regulation, upper computer monitoring, and management. The backbone transmission network of the system is a 100 Mbps industrial Ethernet that supports IEEE802.3 protocol and standard TCP/IP protocol. An industrial special-purpose control local area network is also acceptable. Such a control network is characterized by certainty and repeatability and is capable of I/O sharing, which allows it to achieve fast transmission and real-time control of data.

7.9.2 Technical Characteristics

(1) The fast wet solid-state chemical pretreatment process characterized by low cost and atmospheric temperature will significantly improve the anaerobic digestion of straws, and solve the problems of difficulty in the anaerobic digestion of straws and low gas production rate. After chemical treatment, the gas production from

straws increases by 50–120% as compared with the untreated straws. Solid-state chemical pretreatment will not generate any liquid waste and cause no environmental issues. It happens in atmospheric temperature, and is a simple process with low cost.

(2) In near-synchronous synergistic fermentation of mixed materials, multiple mixed materials are used as the raw material for anaerobic digestion. Before anaerobic fermentation, the raw materials with various properties experience pretreatment through physical and chemical processes, so as to shorten the fermentation period or make the fermentation period controllable and achieve synchronous or near-synchronous fermentation of mixed materials. The anaerobic fermentation process also allows the materials to complement each other's advantages and to achieve synergetic action.

(3) Biogas purification by pressurized water washing allows biogas to be purified efficiently and increases the methane content in biogas to increase by 96%, to generate the fuel gas that meets the international standard for automotive fuel. The purification process only uses water, and all water can be used cyclically, with no pollution. It is the most eco-friendly purification process.

(4) This process realizes a genuine ecological cycle and efficient utilization. Biogas produced by anaerobic fermentation is purified to extract CH_4 which is injected into the natural gas pipe networks or used for vehicles. The biogas residue generated by biogas production from straws is solid and can be directly used as organic fertilizer. Alternatively, it can be made into an organic compound fertilizer that meets the needs of various crops. A part of the biogas slurry is recycled, and the other part is used in vegetable greenhouses or cropland. This cleaner production process completely meets the requirements of the circular economy. Compared with the biogas production process that uses livestock excrement as raw material, this process completely solves the problem of difficult processing and utilization of biogas residue and biogas slurry and causes no secondary pollution. Compared with the straw pyrolysis and gasification process, the straw biogas production process does not generate tar, wastewater, waste gas, or any other pollutant, and generates biogas with high calorific value and high grade, which makes it an eco-friendly biogas production process.

Open Access This chapter is licensed under the terms of the Creative Commons Attribution-NonCommercial-NoDerivatives 4.0 International License (http://creativecommons.org/licenses/by-nc-nd/4.0/), which permits any noncommercial use, sharing, distribution and reproduction in any medium or format, as long as you give appropriate credit to the original author(s) and the source, provide a link to the Creative Commons license and indicate if you modified the licensed material. You do not have permission under this license to share adapted material derived from this chapter or parts of it.

The images or other third party material in this chapter are included in the chapter's Creative Commons license, unless indicated otherwise in a credit line to the material. If material is not included in the chapter's Creative Commons license and your intended use is not permitted by statutory regulation or exceeds the permitted use, you will need to obtain permission directly from the copyright holder.

References

1. Wang Z K. New Trends in China's Population Mobility and Regional Economic Coordinated Development [J]. People's Tribune, 2023 (8): 54–59.
2. Wang J Y. Trend Characteristics of Real Estate Transformation in Typical Countries after Urbanization Rate Reaches 60%[J]. China Development Observation, 2022 (1): 98–103.
3. Ministry of Housing and Urban-Rural Development. Notice of the Ministry of Housing and Urban-Rural Development on Preventing Large-Scale Demolition and Reconstruction in the Implementation of Urban Renewal Actions: Jianke [2021] No. 63 [S]. Beijing: Ministry of Housing and Urban-Rural Development, 2021.
4. Sharma A. Saxena A, Sethi M, et al. Life cycle assessment of buildings: A review[J]. Renewable and Sustainable Energy Reviews, 2011, 15(1): 871–875.
5. Cabeza L F. Rincón L, Vilariño V, et al. Life cycle assessment (LCA) and life cycle energy analysis (LCEA) of buildings and the building sector: A review[J]. Renewable and Sustainable Energy Reviews, 2014, 29: 394–416.
6. Ramesh T, Prakash R, Shukla K K. Life cycle energy analysis of buildings: An overview [J]. Energy and buildings. 2010, 42(10): 1592–1600.
7. Yang Z Y, Hu S, Xu T H, et al. Method and application of global building operation energy use and carbon emissions comparison in the context of carbon neutrality [J]. Climate Change Research, 2023, 19 (6): 749–760.
8. Satish Kumar. Estimating India's commercial building stock to address the energy data challenge [J]. Building Research & Information, 2019, 47: 24–37.
9. Looney B. BP statistical review of world energy [R]. BP Statistical Review: London, UK, 2020.
10. Breeze P. Solar Power[C]. Power Generation Technologies, 2019.
11. Awan A B, Alghassab M, Zubair M, et al. Comparative analysis of ground-mounted vs. rooftop photovoltaic systems optimized for interrow distance between Parallel Arrays [J]. Energies, 2020, 13(14): 3639.
12. A J-W. PV Status Report 2019[R]. Publications Office of the European Union: Luxembourg, 2019.
13. Ahn H K, Park N. Deep RNN-based photovoltaic power short-term forecast using power IoT sensors[J]. Energies, 2021, 14(2): 436.
14. Paydar MAJSC, Society. Optimum design of building integrated PV module as a movable shading device [J]. Sustainable Cities and Society, 2020, 62: 102368.
15. Salimzadeh N. Vahdatikhaki F. Hammad A. Parametric modeling and surface-specific sensitivity analysis of PV module layout on building skin using BIMJJ. Energy Buildings, 2020, 216: 109953.
16. Mboumboue E, Njomo D J R, Reviews S E. Potential contribution of renewables to the improvement of living conditions of poor rural households in developing countries: Cameroon's case study[J]. Renewable & Sustainable Energy Reviews, 2016, 61: 266–279.

17. Nasir M, Khan H A, Hussain A, et al. Solar PV-based scalable DC microgrid for rural electrification in developing regions[J]. IEEE Transactions on Sustainable Energy, 2017, 9(1): 390–399.
18. Ronneberger O, Fischer P, Brox T. U-net: Convolutional Networks for Biomedical Image Segmentation[C]// Medical Image Computing and Computer-Assisted Intervention - MICCAI 2015: 18th International Conference, Munich, Germany, October 5–9, 2015.
19. Zou B, Luo B. Rural household energy consumption characteristics and determinants in China [J]. Energy, 2019, 182: 814–823.
20. Yang Y, Yifang L, Wei Z. Energy consumption in rural China: Analysis of rural living energy in Beijing[J], IOP Conference Series: Earth and Environmental Science, 2017, 81(1): 012063.
21. Reid W V, Ali M K, Field C B. The future of bioenergy[J], Glob Chang Biol, 2020, 26 (1): 274–286.
22. Glob Chang Biol, Calderon. Bioenergy Europe Statistical Report [R], 2020, 26 (1): 274–286.
23. Sun P, Wang Q F, Yi J W, et al. Extraction of Wheat and Rapeseed Planting Areas Based on Decision Tree Classification Method[J]. Agriculture and Technology, 2022, 42 (24): 7–11.
24. Wu Y H. China Hydropower Yearbook[M]. Beijing: China Electric Power Press, 2020.
25. Xu GY, Chen C, Liu DX, et al. Four paths for promoting the green transformation of small hydropower [J]. Water Resources Development Research, 2022, 22 (8): 20–23.
26. Jiang Y. PSDF (photovoltaic, storage, DC, flexible)—A new type of building power distribution system for zero-carbon power systems [J]. HV&AC, 2021, 51 (10): 1–12.
27. Department of Automotive and Tractor of Agricultural Machinery. "Dian Niu-33" and "Dian Niu-55" Electric Tractors[J]. Journal of Northeast Agricultural College, 1960 (3): 1–11.
28. Zhao J H, Xu L Y, Liu E Z, et al. Design for Drive System of Extended—range Electric Tractor [J]. Journal of Agricultural Mechanization Research, 201840 (11): 236–240.
29. Yang K, Zhang D F, Wang P, et al. Design of Braking Control System for Unmanned Electric Tractor[J]. Agricultural Development and Equipment, 2019 (5): 114–116.
30. Tao J H. China Yto Released the First 5G+ Hydrogen Fuel Electric Tractor in China[J]. Contemporary Agricultural Machinery, 2020 (7): 29.
31. Pang K L, Zhang K S, Ma S, et al. Spatiotemporal distribution of activity and its emissions for agricultural equipment in China [J].Acta Scientiae Circumstantiae, 2021, 41 (10): 4268–4279.

The manufacturer's authorised representative in the EU is Springer Nature Customer Service Centre GmbH, Europaplatz 3, 69115 Heidelberg, Germany. If you have any concerns regarding our products, please contact ProductSafety@springernature.com

Printed and bound by CPI Group (UK) Ltd, Croydon, CR0 4YY

26/03/2026

02078916-0006